Gasification Technology

Manoj Kumar Jena • Hari B. Vuthaluru

Gasification Technology

Transforming Waste Into Clean Energy

 Springer

Manoj Kumar Jena [ID]
Department of Chemical and Environmental
Engineering
RMIT University
Melbourne, VIC, Australia

Hari B. Vuthaluru [ID]
Western Australia School of Mines:
Minerals, Energy and Chemical Engineering
Curtin University
Perth, WA, Australia

ISBN 978-3-031-71046-9   ISBN 978-3-031-71044-5   (eBook)
https://doi.org/10.1007/978-3-031-71044-5

This Springer imprint is published by the registered company Springer Nature Switzerland AG
The registered company address is: Gewerbestrasse 11, 6330 Cham, Switzerland

If disposing of this product, please recycle the paper.

This book is dedicated to my father (Late Shri. Pandab Chandra Jena), my mother (Smt. Pankajini Jena), my wife (Dr. Lipeeka Rout), and my affectionate family members.

Preface

With the continuous increase in the population and the achievement of the target in demand for energy, the efficient utilization of fuel for energy generation is essential. Controlled combustion of fuel, known as gasification technology, is one of the prime and old processes of the various strategies adopted for energy generation. This leads to thermochemical conversions of carbonaceous material to valuable products and energy. However, with the continuous growth in science and technology, the industrialization, commercialization, and movement of gasification technology toward sustainability have achieved a more significant milestone, among others. This book is broadly subjected to present the progress in gasification technology from past to present and the expected visions of gasification technology in the future sustainable energy world. This book covers the introduction to gasification technology by explaining the classification and commercialization and associated problems over the last few decades. In accordance with growing interests among interdisciplinary scientific and industrial communities toward gasification, the fundamentals of gasification technology were discussed thoroughly, focusing on the feedstock, product, gasifying environment, and classification of gasifiers in recent times compared to the past. In relation to the challenges of understanding the gasification process for advancement, engineering insights into the gasification mechanisms, including the overall process from particle to reactor level, are discussed rigorously. This will provide a knowledge base for different communities, i.e., industries, research, and academic institutes. In addition, based on the future scope for potential enhancements in industrial application and commercialization, progress has been made for gasification reactors, different aspects of processes were analyzed, and ongoing problems with recent accroaches are also clearly discussed. Moreover, this book also elaborates on the sustainability and prospects of gasification technology, modernization, state of the art in industrial application, and fates and roles of gasification technology in the near future. The key features of this book that readers will find more valuable are the following:

1. Gasification technology is one of the prime technologies that attracts a wide range of feedstocks for efficient and economical production of valuable products.
2. Specific chapters of this book cover the introduction and fundamental understanding of gasification technology, including aspects of classification in the gasifying environment and feedstock, including its properties.
3. The book chapters on engineering insights into gasification technology will present an in-depth understanding of gasification technology from a perspective of chemical reaction engineering, thermodynamics, and transport phenomena.
4. This book details the approaches adopted in the design of gasification reactors and process units, focusing on the gasifying units and reactor models. It also covers the present state of the art of gasification technology, sustainability, and future prospects.
5. This book will provide value-added knowledge and a platform for industries, research, and academic institutes to form a bridge between them for further advancement of gasification technology.

Melbourne, VIC, Australia
Perth, WA, Australia

Manoj Kumar Jena
Hari B. Vuthaluru

Acknowledgments

This book could not have been realized and published without the support of many people who have directly or indirectly been involved in its preparation process. First, I want to give a big thanks to my family for their support throughout every phase of the writing and publication process.

I would also like to thank my friends and colleagues for their invaluable support. I would like to express my sincere gratitude to **Professor Hari B. Vuthaluru**, Curtin University, Perth, Australia, and **Professor Vineet Kumar**, Indian Institute of Technology (Indian School of Mines) Dhanbad, India), who initiated the idea of publishing this book. I would also like to thank **Professor Kalpit Shah,** RMIT University, Melbourne, Australia, for his strong support and professional guidance. I am also very grateful for their constant encouragement and continuous support throughout the preparation of this book. It has been a privilege for me to work with amazing people like them. I would also like to convey my sincere gratitude to the authors across the globe who have worked in this area to produce the relevant papers/journal articles which were of immense help in writing of this book. We want to thank all the authors from different journal articles and publishers for allowing the use of the figures and plots in this book with appropriate copyright permissions.

Introduction

In recent years, the development of energy technology has become essential to balance the demand for energy and other needs. Among all kinds of technologies, the technologies relevant to the thermochemical conversion of solid fuel through the gasification process have been widely accepted as cost-effective and very efficient. For several decades, the gasification technology has been used for power as well as generating value-added syngas products. From the source knowledge, it was stated that solid fossil fuel was primarily used for it, and earlier developed gasification technology was called coal gasification technology. However, continuous growth, technology, and advanced characterizing techniques have led to many renewable solid fuel sources that can substitute solid fossil fuels. In general, the past gasification process and the intended purpose of using it in industries have been significantly changed in a positive manner. The efforts that have been carried out in the present development of gasification vary far from the past. This indicates that gasification technology by itself has the potential to be widely accepted as a diversified feedstock (solid fuel) flexibility and can be applied to many different industrial processes. In principle, the advancement of gasification technology is worthwhile for future generations. This will resolve problems associated with energy and the formation of valuable chemicals to create a platform or landmark. However, to make an advanced approach to gasification technology, a set of knowledge base should be acquired from a fundamental understanding of gasification mechanisms, reaction kinetics, and design of gasification reactors. It is believed that the book *Gasification Technology: Past, Present and Future* by Dr. Manoj Kumar Jena and Professor Hari B. Vuthaluru attempts to provide these insights into gasification technology.

Thus, it gives us great pleasure to contribute this foreword. Before writing it, we reviewed the book chapters and details of the contents. We were satisfied with the book's structure and the topics covered. The book is divided into six chapters. The first chapter deals with the introduction of gasification technology. The chapter also includes the history of gasification and the type of gasification technology, as well as the challenges associated with past gasification technologies. The second chapter focuses on the fundamentals of gasification technology, such as the properties and composition of feedstocks, products of gasification technology, gasification environment, and classifications of gasifiers. The third chapter of this book covers engineering insights into gasification technology, in particular reaction kinetics, catalysis, gasification thermodynamics, and transport phenomena such as heat, mass, and momentum transfer processes. The fourth chapter discusses the gasification reactor and processes, which includes the classification of modern gasification reactors, the model of the reactor, and gasification systems, including major problems and approaches related to gasification technology. The fifth chapter discusses the sustainability and prospects of gasification technology, covering where the advancement of gasification technology, state of the art in industrial application, fates and roles of gasification technology in the next generation, and the role and responsibilities of scientific communities for gasification technology protection are discussed comprehensively. The sixth chapter summarizes the major conclusions and recommendations for future work for advancing gasification technology.

This book presents all important data supported by appropriate figures and tables to help readers understand basic concepts. This became one of the most important features of this book when we began reading its contents. Thus, we believe this is a very important book for Springer Nature, and it will be very beneficial for students, researchers, scientists, and faculty members to conceptualize this area of research across the globe.

Melbourne, VIC, Australia   Manoj Kumar Jena
Perth, WA, Australia   Hari B. Vuthaluru

Contents

Chapter 1
Introduction

Gasification: A Legacy of Transformation and Innovation from Nineteenth-Century Coal to Twentieth-Century Clean Energy

1.1 History of Gasification

In the last few decades, around the beginning of the eighteenth century, a scarcity of energy in the world has led us to think of alternative ways of getting energy due to the consistent rise in the human population on the earth. More efforts were needed to eradicate those problems, and at the same time, it was also focused on optimizing the use of oil and natural gas resources. Simultaneously, it was noticed that people from different rural areas started using the dung obtained from the pet animals for cooking purposes [1]. They usually make palm plates using this dung and dry them in regular sunlight for a couple of days. Once it gets dried completely, they use it as fuel to generate heat for cooking. Apparently, by looking at that kind of behavior, a further thought was raised in the mind of the researcher/scientist/engineer. They found that solid fuel material being prepared from dung might get combusted entirely and that the combustion of solid burns off all carbon, including the organic and inorganic compounds. Only heat energy can be extracted from the process of combustion of solid carbon material. However, the requirement for energy cannot be only overcome by heat energy; the formation of a potential product could lead to the fulfillment of the energy demand through the formation of valuable chemicals [2]. After further effort was made, the advancement of a process called combustion led to the development of the gasification process. In an overall approach, the invention of the gasification process was generated by the demand for energy from people that provoked them to adopt different ways of generating energy from available resources. Therefore, the requirement or need for something was fulfilled by creating something new from existing methodologies. The formation of the gasification process is equally credited to engineers, scientists, and general humans worldwide.

© The Author(s), under exclusive license to Springer Nature Switzerland AG 2024

M. K. Jena, H. B. Vuthaluru, *Gasification Technology*,

https://doi.org/10.1007/978-3-031-71044-5_1

1.1.1 What Is Gasification?

From a scientific and engineering point of view, the terminology "gasification" can be referred to as science as the so-called gasification process and can be as gasification technology from an engineering point of view [3]. The objective of gasification is to efficiently carry out partial combustion (which can be stated as partial oxidation) of carbonaceous material so it could lead to more CO, CH_4, and H_2 in contrast to combustion with the formation of a rich amount of CO_2. The primary objectives mainly control the adoption of the gasification process; if the formation of CO and CH_4 is the prime one, the gasification with O_2 or CO_2 can be a significant path to achieve this [4]. If H_2 is the primary requirement for formation, gasification with H_2O could be the potential route to gain the objective, highlighting the promising potential of gasification for future energy production and environmental sustainability.

In addition, the approaches to advancing gasification from science and engineering to being a gasification process and gasification technology are interrelated. Those interdependent approaches are mainly focused on material, formation/breaking of bonds, and formation of different compounds/molecules from the science perspective; however, the design of the reactor, controlling of reaction kinetics, and the feasibility of the gasification process are determined from engineering side with enhancing the process parameters is called as gasification technology [5].

1.1.2 Gasification Benefits

The development of gasification processes and technology has many applications in several industrial applications, such as gasification of carbon material, coal, biomass, biosolids, and other carbonaceous materials. Also, gasification technology has been implemented in industries as an integral part of other processes to intensify overall processes. This helps to increase revenue by scaling up the minor process to the industrial level of oil, natural gas, and other energy industries [6, 7]. From the material point of view, the benefit of gasification lies in the efficient utilization of coal (which was used at the beginning) for energy generation and value-added products with low emissions of CO_2. As high CO_2 emissions lead to an increase in green gas, adversely affecting the human environment, gasification technology is considered to be a clean technology. Therefore, the overall gasification process and technology development are dedicated to industries' processes and intensification. It also dramatically balances the ecosystem by emitting less dangerous gases to the environment, such as NOx, SOx, CO_2, and particulate matter (generated from inorganic pollutants). In conclusion, adding gasification to the existing process is a cost-effective, energy-efficient, and eco-friendly approach for the overall process. A detailed explanation of this will be elaborated in the following chapters.

1.2 Broad Classification of Gasification Technology

It was broadly classified into three major types based on using coal as solid carbon material for the gasification process to diversify the gasification technology and widen its application. This is because coal was first used for the gasification process than other applications. Also, the extracted material from the mine has different values as some coal is obtained in the form of slurry, and some may be in the form of a big chunk as received from other places. Those as-received coal samples may need further processing before the gasification process to optimize process operation. The gasification technology was classified into three major types, i.e., ECUST opposed-multi-burner (OMB) coal-water-slurry (CWS) gasification technology, GE Texaco CWS gasification technology, and Shell pulverized coal gasification technology. Hence, the broad gasification of existing technologies is provided here based on the coal gasification technology developed by different companies/industries.

1.2.1 ECUST Opposed-Multi-Burner (OMB) Coal-Water-Slurry (CWS) Gasification Technology

ECUST-opposed multi-burner coal-water-slurry gasification technology is one of the critical technologies developed for developing coal-based chemistry products, coal-based clean fuel, advanced IGCC power generation, poly-generation systems, hydrogen production, fuel cells, etc. The name of this gasifier came from the history of collaboration between the East China University of Science and Technology (ECUST) and the Yankuang group in accomplishing the OMB CWS coal gasification technology project during "the 9th Five-Year Plan." After 5 years of completion of that project with continuous development and industrial application, the OMB CWS gasification technology became stable and mature in addition to other technologies for process intensification [8–10]. 144 OMB CWS gasifiers have been employed in 52 countries, including China, in a scale-up manner with integration into other technologies. Figure 1.1 shows a schematic view of the OMB CWS gasification processes.

The OMB CWS gasification technology was first demonstrated with a capacity of 1000 TPD (tons per day) in 2005. It was enhanced further to a 3000 TPD OMB CWS gasifier to become the largest gasifier in the world at that time. The successful operation of the demonstration unit jointly operated by Yankuang Cathay Coal Co., Ltd. (1150TPD, 4.0 MPa) and Shandong Hualu Hengsheng Chemical Co., Ltd. (750TPD, 6.5 MPa) suggests the feasibility of imported CWS gasification technology and stable with OMB gasifier [12]. A detailed list of OMB gasification technologies and their respective industrial application is shown in Table 1.1.

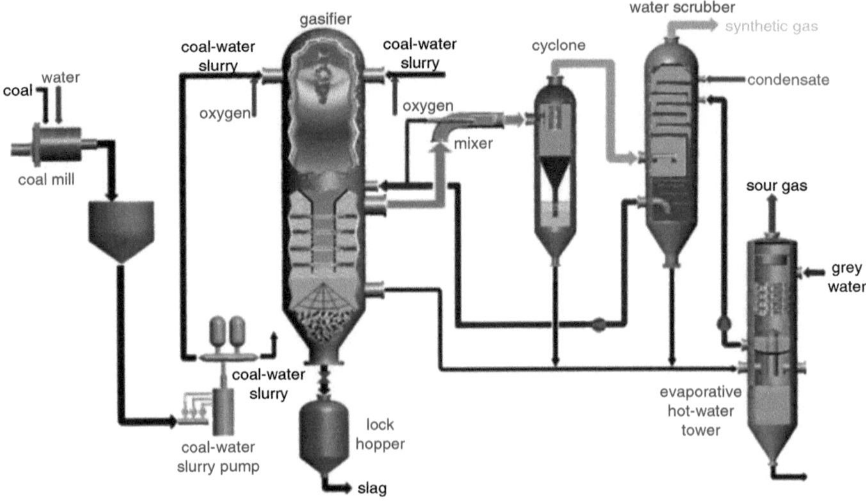

Fig. 1.1 Flow chart of the opposed multi-burner (OMB) coal water slurry (CWS) gasification system [11]

Table 1.1 List of OMB gasification technologies and their respective industrial applications [13]

Project name	Pressure MPa (G)	Trains	Product
Hualu Hengsheng Chemicals Co., Ltd.	6.5	1	MeOH, NH$_3$
Yankuang Cathay Coal Chemicals Co., Ltd.	4.0	3 (2 + 1)	MeOH, power
Yankuang Luhua	4.0	1	NH$_3$
Ningbo Wanhua Co., Ltd	6.5	3 (2 + 1)	MeOH, CO, H$_2$
Henan Xinlianxin Fertiliser Co., Ltd.	6.5	3 (2 + 1)	NH$_3$
Ningbo Zhongjin Petrochemical Co. Ltd.	1.5	2 (1 + 1)	Fuel gas
Zhongyan Kunshan	6.5	2 (1 + 1)	NH$_3$
Jiangsu Sanmu	4.0	2 (1 + 1)	Octanol
Yihua Chemical Industry Co., Ltd.	6.5	3 (2 + 1)	NH$_3$
Korea TENT	4.2	2 (1 + 1)	Syngas

1.2.2 GE Texaco CWS Gasification Technology

The history says that the Texaco gasification technology was initially developed in 1940 using natural gas as feedstock. The Texaco gasification process was commercialized in 1950. However, the liquified feed was further developed in 1956. Subsequently, this process was a massive success during the oil shortage period. The actual need currently being used came after GE acquired it from Texaco; it is now completely called GE Texaco CWS gasification technology [14].

After that, the first 15 TPD (tons per day) of pilot scale gasifier was established in Montebello, CA, in the U.S. Considering the operation of the Texaco gasification process, some of the disadvantages led to limiting its application in fields/industries.

Those disadvantages include (i) no development was done on slurry additive during that period, causing the use of low concentrations of coal for operation, (ii) drying and mechanical operation such as pulverization of feedstock leading to the process to become more complicated, and (iii) the abrasive nature of slurry during evaporation leading to may fouling issues and (iv) loss of coal during vaporization processes. In addition to the above, the GE gasification with coal water slurry preparation led to the development of an entrained flow slagging gasifier. In contrast to the previously developed Texaco gasification operation, the GE operation allows the use of a radiant boiler and a total quench for the syngas cooling concept [15, 16]. A schematic view of the Texaco gasification process is presented in Figs. 1.2 and 1.3.

1.2.3 Shell Pulverized Coal Gasification Technology

The Shell pulverized coal gasification technology (as schematically shown in Fig. 1.4) can cover a wide range of feedstocks, i.e., coal, in forming clean syngas efficiently to generate electricity in an ICGCC plant. Creating high-quality syngas has already been demonstrated for low-rank lignite coal, high-rank bituminous coal, and petroleum coke. In 1993, in the Netherlands, shell technology was developed by Denmark in partnership with the Dutch Electricity Generating Board (N.V. Sep) [18]. Over the period, followed by 3 years of demonstration, the plant became a part of the Dutch electricity generating system. The shell gasification technology system

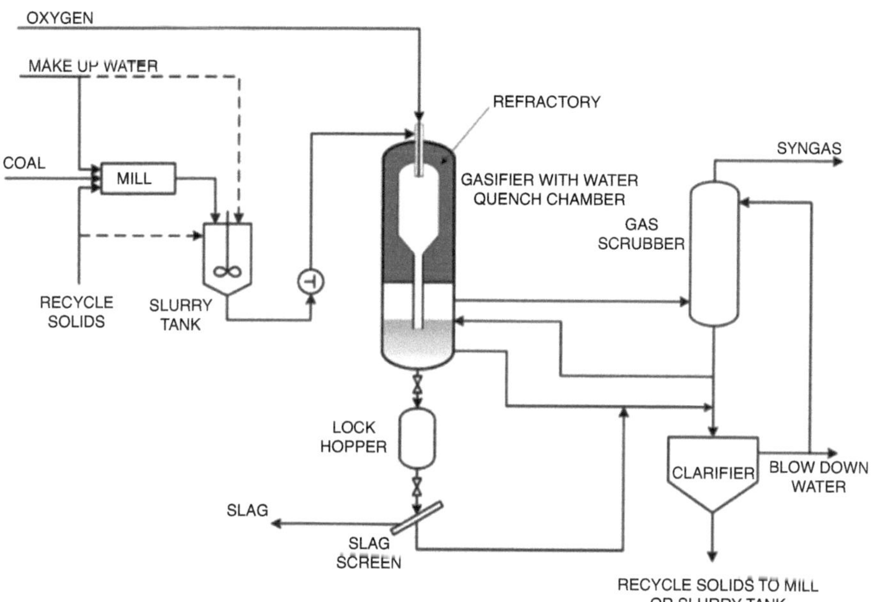

Fig. 1.2 Schematic of a typical GE coal gasification system with a quench chamber [17]

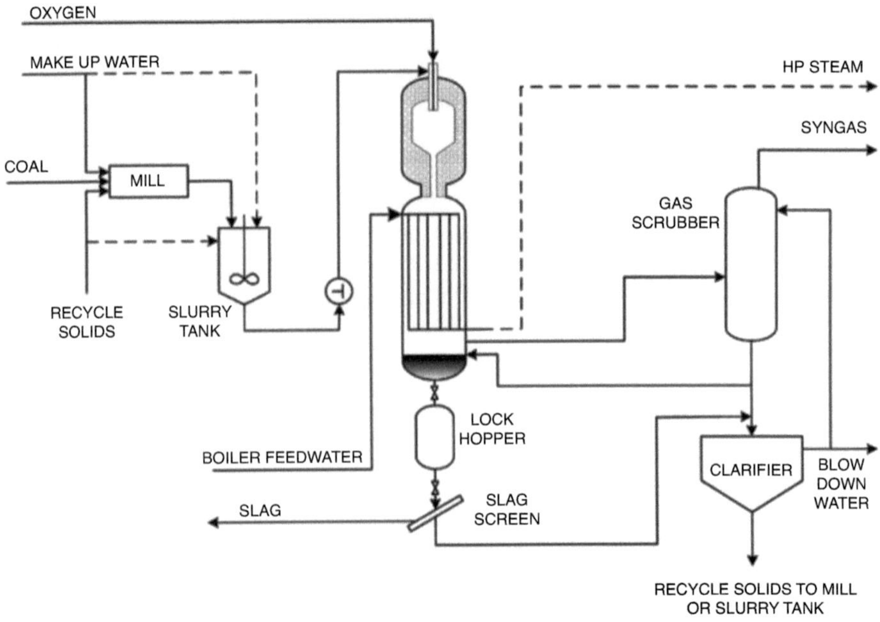

Fig. 1.3 Schematic of a typical GE gasification system with radiant syngas cooler [17]

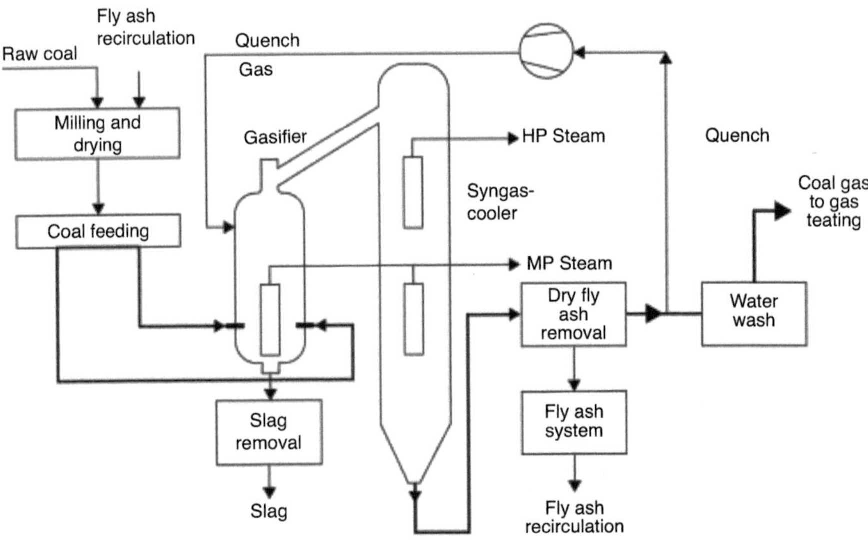

Fig. 1.4 Schematic view of Shell gasification technology using convective syngas cooling [11]

was developed for gaseous and liquid feedstock, coal, lignite, and petroleum coke. All over the world, shell coal gasifiers are primarily attracted to countries like Australia, Southern Africa, Indonesia, Europe, and North America. In addition, the current focus was on the marketing and transportation of the slurry pipeline. In the recent advancement of coal gasification technology, the Shell gasification process led to the implementation of a dry-feeder, oxygen blown to the hydrodynamic entrained flow gasification process. That results in the enhancement of clean syn gas. In the pulverized coal gasification process, the coal was fed through a dynamically opposed burner, reacting with a gasifying agent at an operating temperature. The gasification operation temperature was maintained to ensure there would not be any coal in molten condition to enable the flow of gas smoothly down the gasifier wall and come out through the slag trap [19, 20].

1.3 Commercialization of Gasification Technology

Considering the popularity of gasification technology, many were developed by different technology providers. However, only some of them were commercialized successfully compared to others, and those are listed and briefly described below.

1.3.1 Lurgi Gasifier

The Lurgi gasifier is based on the Lurgi dry-ash gasification technology, which has been developed in the early 1930s. However, the first commercialized plant was established in 1936. In the subsequent decade (in 1950), Lurgi and Ruhrgs jointly developed a technology for bituminous coal apart from traditional lignite coal feedstocks [21]. The Lurgi gasifier is operated in a pressurized condition with a dry ash moving bed continuously producing syngas. The gasification operation is usually carried out at low temperatures, and a high ratio of steam to oxygen was used to equilibrate the temperature to moderate so that the ash usually comes out as dry ash without any melting conditions [22, 23]. A schematic diagram of the Lurgi gasifier is shown below in Fig. 1.5.

1.3.2 GE Energy Gasifier

The GE energy gasifier was initially developed with the name ChevronTexaco or Texaco gasifiers but was renamed to GE Energy gasifier. GE acquired its technology in 2004, and the business acquisition occurred in 2019. Prior to this, however, it was successfully commercialized over 45 years using a variety of feedstocks, including natural gas, heavy oil, coal, and petcock [1]. In addition, the coal gasification

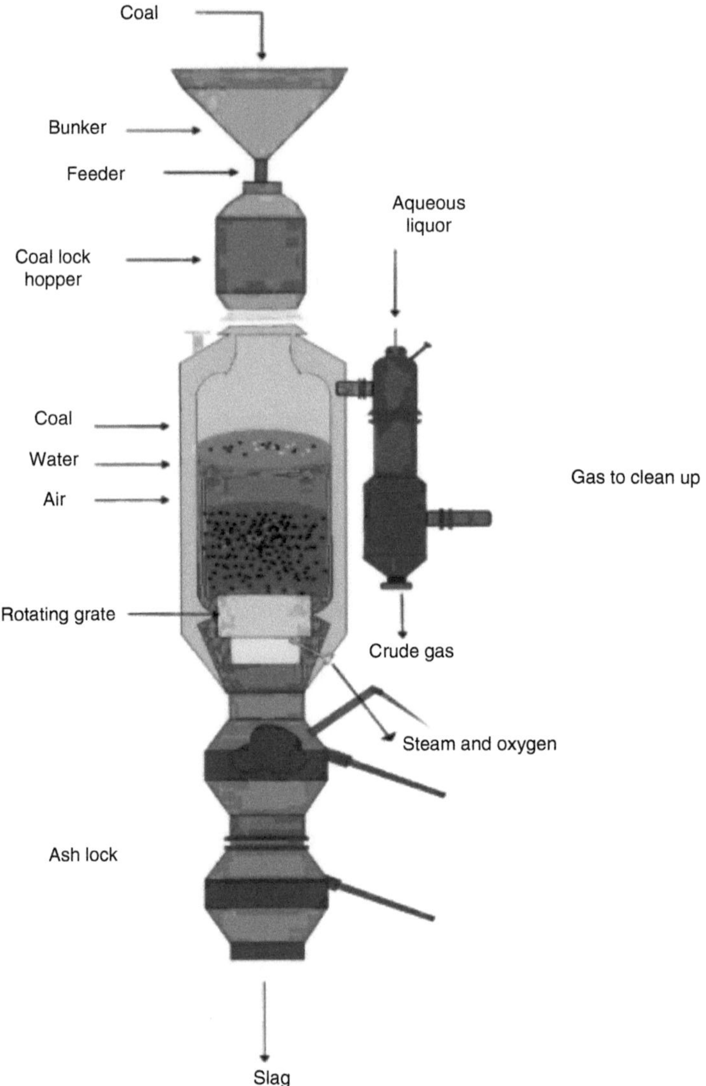

Fig. 1.5 Diagram of the Lurgi dry-ash gasifier. (Source: Sasol) [24]

operation was developed for coal based on the initial observations made from oil and natural gas gasification. The earlier developed work occurred at 15 TPD (tons per day) in Montebello, California, and subsequently, it was moved to China. This is usually operated with a single-stage downward-feed, entrained flow gasifier. The syngas is produced in a refractory-lined reactor from coal/water slurry (~ 65% in wt.) and in an environment of >95% oxygen [18]. A schematic view of the GE energy gasifier is shown in Fig. 1.6.

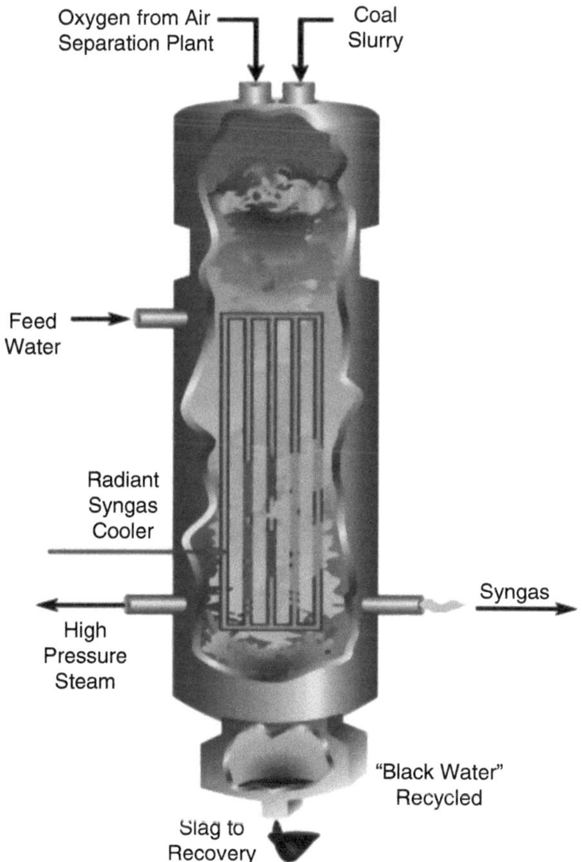

Fig. 1.6   GE energy gasifier [1]

1.3.3   Shell Coal Gasifier

The Shell coal gasification process (SCGP) was initially developed from the Shell gasification process, with an observation from partial oxidation of heavy oils and natural gas. After that, the feedstock used for the gasification process was expanded to a wide range of solid feedstock, including coal [25]. The initial commercialized gasifier by Shell Internationale Petroleum Maatschappij (SIPM) had a capacity of 6 TPD (tons per day) in Amsterdam in 1972. Shell's coal gasification technology uses a dry feed in pressurized entrained flow conditions. The dried pulverized coal was fed to the gasification reactor through lock hoppers using carrier gas as nitrogen and syngas. The coal reacts with preheated 95% pure oxygen and steam as moderator at 2700–2900 °F. In this Shell gasification process, a small amount of carbon dioxide is released with syn gas without any hydrocarbon liquids or gases [22]. A schematic view is presented in Fig. 1.7.

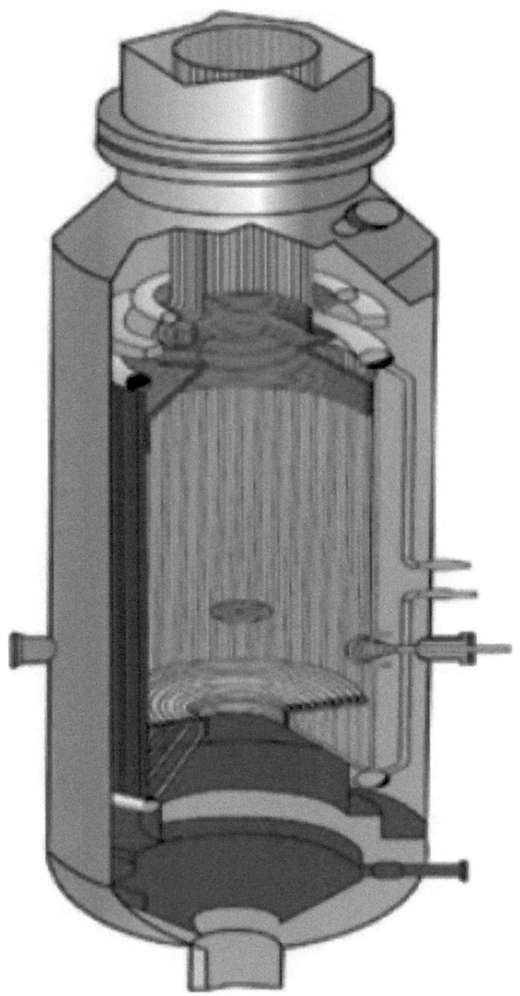

Fig. 1.7 Shell coal gasifier. (Source: Shell) [1]

1.3.4 ConocoPhillips E-Gas Gasifier

Global Energy, Destec, and Dow initially owned the ConocoPhilips E-Gas gasification technology, which ConocoPhilips then took. Later or currently owned by CB&I, E-Gas™ coal gasifier technology [26]. This technology was first demonstrated at Louisiana Gasification Technology Inc. (LGTI) with an integrated gasification combined cycle (IGCC) plant. The E-GasTM coal gasifier operates in a high-pressurized entrained flow gasifier with a unique two-stage operation [27]. The gasification operation usually involves a highly exothermic oxidation reaction that typically runs at 2600 °F and leads to a formation condition that does not allow the formation of hydrocarbon gases and liquids. As produced, the residual ash finally exits through the hole at the bottom of the gasifier to the water quench, as shown in the schematic view of the CoconoPhillips gasifier in Fig. 1.8.

Fig. 1.8 Schematic view of CoconoPhillips gasifier [1]

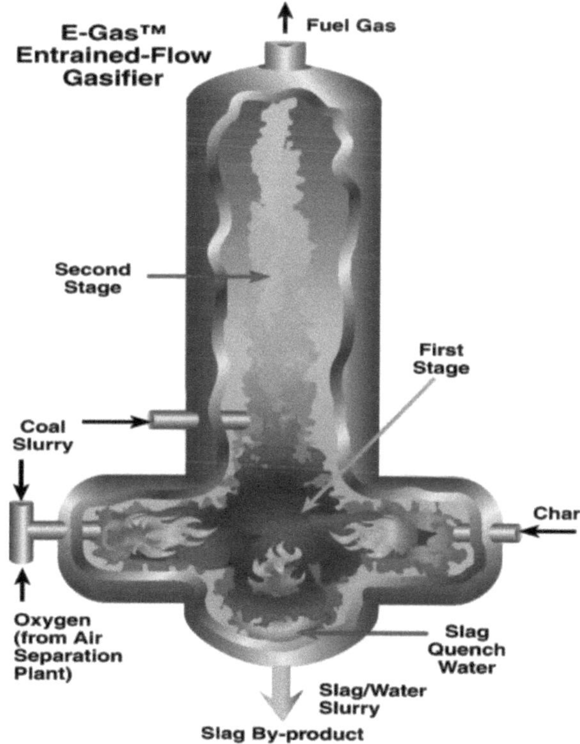

1.3.5 Kellogg Brown and Root (KBR) Gasifier

The KBR gasifier, also known as TRIG™ transport integrated gasifier with the implementation of an advanced circulating fluidized-bed reactor. Figure 1.9 shows a schematic picture of the KBR transport gasifier. The technology was initially trailed at the Power Systems Development Facility (PSDF) in Wilsonville, Alabama. After that, a series of tests and trials were done in 1999 to demonstrate a stable and steady-state operation utilizing a wide range of feedstock [28]. The KBR transport gasifier is the basis for the IGCC plant, which uses air as a gasifying agent for Lignite coal in Kemper Country, Mississippi. In operation, the KBR transport gasifier mixes steam and air/oxygen in the low mixing zone while solid fuel, sorbent, and a few more oxygen and steam are added in the upper mixing zone. The gasifier is usually manufactured with a refractory-lined pipe, which allows for the use of less expensive metal for the reactor cell [29, 30].

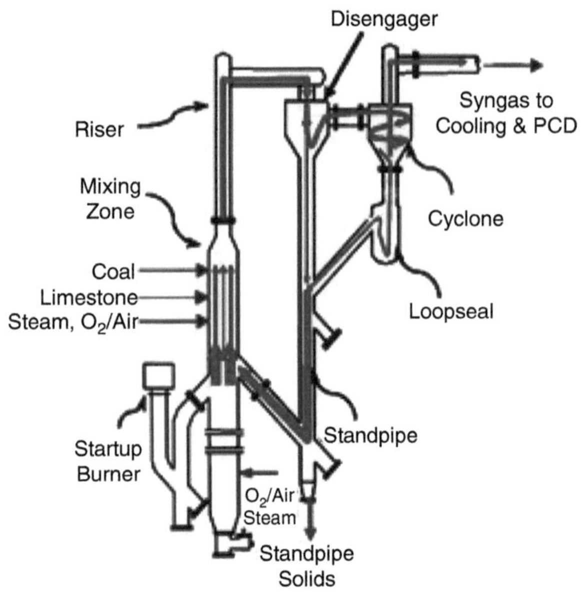

Fig. 1.9 Schematic picture of KBR transport gasifier. (Source: Southern Company) [31, 32]

1.3.6   British Gas/Lurgi (BGL) Gasifier

The British Gas/Lurgi gasifier was adopted from the Lurgi dry ash gasifier. However, further improvement was made considering the production of non-leachable vitreous slag rather than dry ash, improvement of reactor throughput, increase in the standard feed size, decrease in the consumption of steam, and recycling of tar/oils to increase the yield of carbon monoxide and hydrogen in the product gas [33]. This gasification operation is carried out at 3600 °F (~2000 °C). However, during a coal mixture with coarse coal and fine, the feed was introduced from the top to a refractory-lined gasifier through a lock hopper system [34]. The product gas from the gasifier is usually released at 840 °F (450 °C) with an opening at the top of the gasifier vessel, as shown in Fig. 1.10.

1.3.7   MHI Air-Blown Gasifier

Considering the previously developed gasification technology, with a goal of higher gasification efficiency, a dry feed with a unique two-stage reactor was designed, named Mitsubishi Heavy Industries (MHI) gasification technology. The MHI technology in Japan commenced in 1980 with the collaboration of several other power utility industries. The operation mode of the MHI gasifier is similar to that of the E-Gas™ gasifier, with an entrained-flow reactor in addition to a two-stage operation [1]. However, the current research and development focus is on understating and

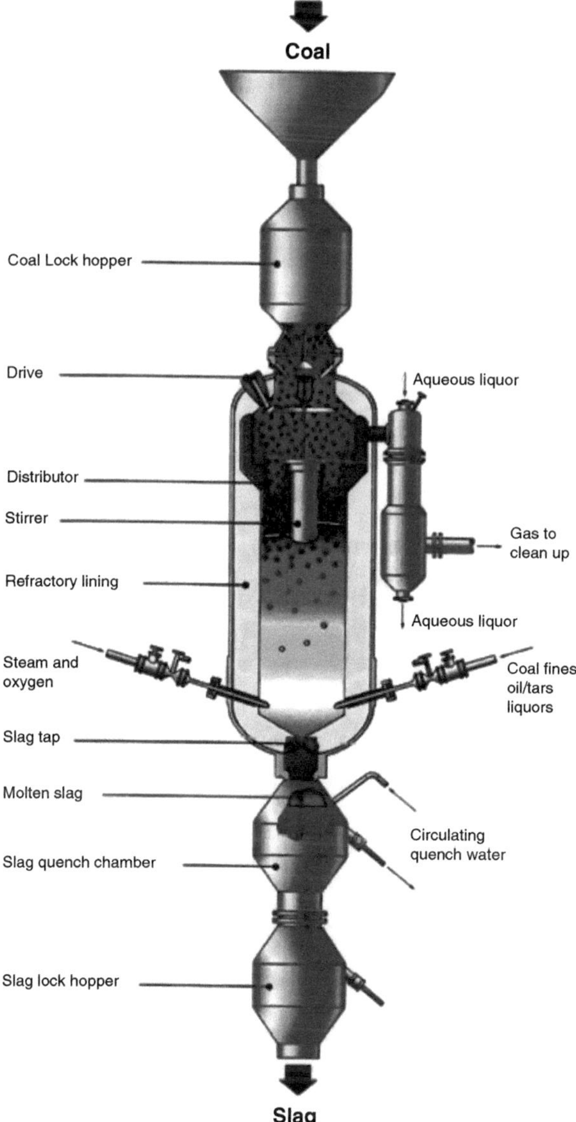

Fig. 1.10 BGL gasifier. (Source: GL Industrial Services, Leicestershire, UK) [35, 36]

developing an oxygen-blown system for forming coal into valuable chemicals. The raw syngas are released from the system at around 2200 °F, which is enough temperature to contain very few hydrocarbon gases and liquids [37], as shown in the schematic view of the MHI gasifier in Fig. 1.11.

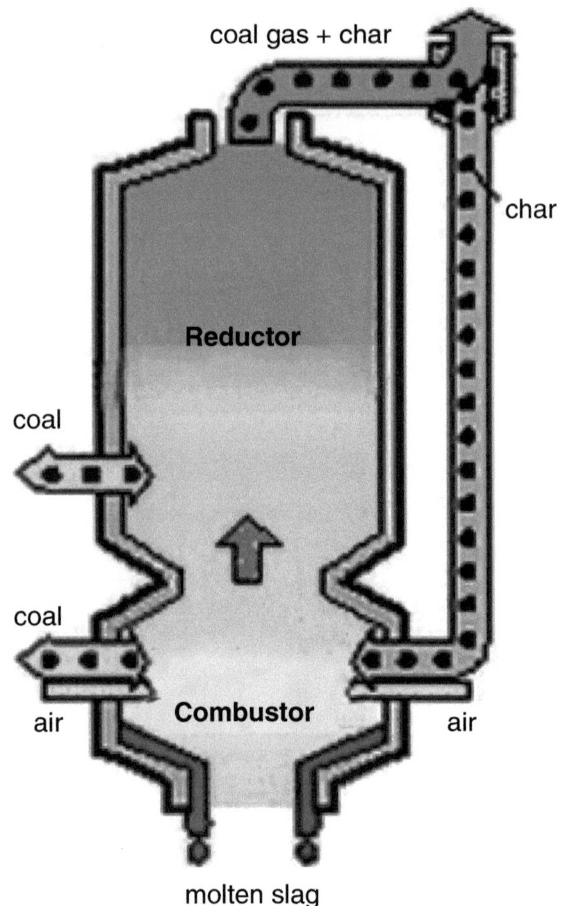

Fig. 1.11 A schematic view of the MHI gasifier. (Source: Mitsubishi Heavy Industries, Ltd.) [38]

1.3.8 Summary of Commercial Processes from Past to Present and Future Outlook

The impact of technology being commercialized depends on several factors, such as the capacity to reach desired product yield, the efficiency of thermal operation, capital cost, and flexibility on scale-up of technology, i.e., from lab scale to pilot scale. The capacity to reach the desired product ratio depends on the type of feedstock, which can be a single one, such as coal, biomass, and other solid carbonaceous materials, or could be combined with any other co-feedstock [39]. Predicting the commercialization of gasification technology in the past primarily depended on the diversification of feedstock from oil and natural gas to coal. Also, it relies on the quality of syngas as received from gasification operations in a continuous system.

However, knowing the impact of char and biochar, the current technologies are being commercialized based on the type of reactor, which could be a fluidized bed reactor, fixed bed, or any other thermal reactor. The reactor selection depends on the type of feedstock used, such as coal, coke, biomass, other plastic materials, and solid waste [40].

In addition, gasification technology can also be commercialized based on other specific conditions, such as availability of feedstock, flexibility of operation, and easier relocation if needed. Overall, there are no criteria for making a technology for commercialization. The current developed technology should be advanced in comparison to existing technology. However, in going further in the future, the development of technology for the future can be expected to be commercialized based on how easily it can be integrated with other parts of the plant with less effect on the existing system, how it can solve the problem which has been encountering from existing technology with less impact on the product quality and operation efficiency. Hence, the future commercialization of gasification technology depends on factors that will fulfill the future energy demand and how those factors are related to newly developed gasification technology [41–43].

1.4 Challenges Associated with Gasification Technology

Regarding the prospects of gasification technology development, the factors involved in scaling up from bench scale to pilot plant scale are associated with many changes. However, those challenges are not only engaged with the engineering aspects but also involve some understanding of the fundamental aspects of science. All challenges are broadly classified into four major parts: global technical and scientific challenges for the gasification process, challenges related to the availability of suitable feedstock, gasifying environment optimization, and gasification efficiency enhancement [41, 44]. A brief discussion on each challenge is presented below.

1.4.1 *Global Technical and Scientific Challenges for the Gasification Process*

The continuous progress of the human population and energy demand have attracted people to the formation of more advanced gasification technology, which can be used in a broad range of applications. On the contrary, this advanced technology development needs more understanding of engineering and science. Prioritizing the global technical challenges is majorly associated with developing a suitable reactor that can be operated in a wide range of temperatures, feedstock, and gasifying agents and with larger capacity. On the other hand, there still needs to be more knowledge required pertinent to the fundamental understanding of reaction kinetics for different types of solid fuel in varying gasifying environments. In addition, more knowledge is necessary to understand the hydrodynamic behavior of additional gasifying reactors in hot flow conditions [45, 46].

To overcome those challenges, both fundamental and applied research are being carried out currently in different parts of the world, and this continuous progress in research must help overcome future expectations from scientific communities.

1.4.2   Suitable Feedstock

Feedstock properties and composition play a prominent role in the gasification process. In the gasification process, particularly operation conditions, the availability of suitable feedstock could lead to operation efficiency and quality of product syngas. To operate well, understanding gasification operations that are inadequate for proper feedstock is essential. In another way, estimation says that gasification operating conditions and adequate feedstock are interrelated; lack of any one of these could make the process incomplete. However, in some specific cases, the particle size could play a vital role in the path of product gas formation, as in the case of low-rank coal. It could be an option for optimizing the product gas composition in various gasifying environments. On the contrary, in the case of developing technology, the scaling of technology from bench scale to pilot scale also needs suitable optimization of feedstock with particular feedstock properties. The properties of feedstock vary with the type and rank of material, such as coal, biomass, and other carbonaceous materials like biosolid. The properties differ a lot from each other [47, 48].

Hence, the reason for specific operating conditions is that considering suitable feedstock with the appropriate properties is essential for getting a well-optimized product yield. Also, it is necessary to understand the fundamental aspects to get more insights into advanced gasification technology for a wide range of feedstock, including all solid carbonaceous materials. This could be one of the future concerns for scientific communities, which will need to gather more knowledge on those aspects.

1.4.3   Optimized Gasifying Environments

Considering factors affecting the control mechanism of gasification reactions, a high yield of a particular component in overall gas composition needs optimization of the gasifying environment [49]. For example, if the formation of hydrogen production is required in a gasifying climate, the ratio of steam to oxygen needs to be correctly optimized. This is because most steam gasification reactions occur at high temperatures, primarily above 700 °C. Under this situation, with a mixed environment system, the char-O_2 and char-H_2O reaction mechanisms vary. At 700 °C, the char-O_2 usually occurs in a controlled mass transfer limitation mechanism as the chemical reactivity of char contains the char-H_2O reaction, so-called the kinetic controlled mechanism [50]. This leads to different tools for each, with differing mechanisms for product gas formation. This kind of understanding is not limited to the mixture of steam and oxygen environment but is also applicable to mixing other gasifying agents such as CO_2, H_2, etc.

This leads to the idea that improper optimization of the gasifying environments at particular operating conditions could significantly affect a product's yield. However, a significant knowledge gap is in their use of optimizing the gasifying environment by selecting and obtaining a specific product gas composition. This must be fulfilled to enable us to continuously design or fabricate large-scale reactors for gasification operation in integrated plants [51].

1.4.4   Enhancement of Gasification Efficiency

Many more factors control the gasification efficiency; it is not only limited to the quality of feedstock and gasifying environment but also depends on other factors, including the rate of heat supply, rate of feedstock input in a continuous, and product recovery in a constant, steady state operation, type of gasification reactor, mechanism of hydrodynamic process in a constant bed and other operating conditions such as temperature and pressure [52]. Those factors are equally responsible for enhancing the gasification operation, but optimizing all to bring to a common platform is challenging. However, in the gasification process, one parameter is usually optimized to keep other parameters constant. Though it is possible in theoretical studies, it may or may not be likely to happen in actual operation [42, 53].

In addition to this, in an integrated system, the performance of the gasification system largely depends on the plan attached to the front and back ends and their respective performances. This is because every system largely depends on each other in an integrated approach. For example, if one driver is attached at the front end of the gasification system, the performance of the drier could primarily affect the gasification performance if the gasification operation does not receive the feedstock at the desired dried condition [54]. This suggests that factors that could cause an increase in the efficiency of gasification operation are only limited to the actual system but also depend on other external factors and systems attached to the operation. This is one of the biggest challenges occurring in industries and needs to be studied coherently by researchers in the future.

References

1. Breault, R. W. (2010). Gasification processes old and new: A basic review of the major technologies. *Energies, 3*, 216–240. https://doi.org/10.3390/en3020216
2. Phillips, J. N., Booras, G. S., & Marasigan, J. (2017). The history of integrated gasification combined-cycle power plants.
3. Higman, C. (2011). 5 – Gasification process technology. In M. R. Khan (Ed.), *Advances in clean hydrocarbon fuel processing* (pp. 155–185). Woodhead Publishing.
4. Ahmad, A. A., et al. (2016). Assessing the gasification performance of biomass: A review on biomass gasification process conditions, optimization and economic evaluation. *Renewable and Sustainable Energy Reviews, 53*, 1333–1347.
5. *Chemistry, thermodynamics, and kinetics.* In *Handbook of gasification technology.* (2020). pp. 107–140.
6. Cooke, B. H., & Taylor, M. R. (1993). The environmental benefit of coal gasification using the BGL gasifier. *Fuel, 72*(3), 305–314.
7. You, S., et al. (2016). Comparison of the co-gasification of sewage sludge and food wastes and cost-benefit analysis of gasification- and incineration-based waste treatment schemes. *Bioresource Technology, 218*, 595–605.
8. Shen, Z., et al. (2021). A deep insight on the coal ash-to-slag transformation behavior during the entrained flow gasification process. *Fuel, 289*, 119953.
9. Guo, Q., et al. (2017). *Opposed multi-burner (OMB) coal-water slurry Gasification Technology and its industrial applications.* East China University of Science and Technology. Private communication.
10. Wang, F., Yu, G., & Guo, Q. Development of coal gasification technology in China. World Coal Association. https://www.worldcoal.org/development-coal-gasification-technology-china. Accessed on 30 Nov 2018.
11. Gray, D. D. (2017). 8 – Major gasifiers for IGCC systems. In T. Wang & G. Stiegel (Eds.), *Integrated Gasification Combined Cycle (IGCC) technologies* (pp. 305–355). Woodhead Publishing.
12. Wang, F., et al. (2021). Opposed multi-burner gasification technology: Recent process of fundamental research and industrial application. *Chinese Journal of Chemical Engineering, 35*, 124–142.
13. Gong, Y., et al. (2017). Progress on Opposed Multi-Burner (OMB) coal-water slurry gasification technology and its industrial applications. *Energy Procedia, 142*, 1089–1094.
14. Gray, D. (2017). Major gasifiers for IGCC systems. In *Integrated gasification combined cycle (IGCC) technologies* (pp. 305–355). Elsevier.
15. Liu, K., Cui, Z., & Fletcher, T. H. (2009). Coal gasification. In *Hydrogen and syngas production and purification technologies* (pp. 156–218).
16. Yang, D., Li, S., & He, S. (2024). Zero/negative carbon emission coal and biomass staged co-gasification power generation system via biomass heating. *Applied Energy, 357*, 122469.
17. Higman, C. (2008). Chapter 11 – Gasification. In B. G. Miller & D. A. Tillman (Eds.), *Combustion engineering issues for solid fuel systems* (pp. 423–468). Academic Press.
18. Sánchez-Hervás, J. M., Moya, G. M., & Ortiz-González, I. (2019). Current status of coal gasification. *New Trends in Coal Conversion*, 175–202.
19. Guo, X., et al. (2007). Performance of an entrained-flow gasification technology of pulverized coal in pilot-scale plant. *Fuel Processing Technology, 88*(5), 451–459.
20. Bell, D. A., Towler, B. F., & Fan, M. (2010). *Coal gasification and its applications.* William Andrew.
21. Robson, B. (1977). A review of gasification for power generation. *International Journal of Energy Research, 1*(2), 157–177.
22. Littlewood, K. (1977). Gasification: Theory and application. *Progress in Energy and Combustion Science, 3*(1), 35–71.

23. Bunt, J., Wagner, N., & Waanders, F. (2009). Carbon particle type characterization of the carbon behaviour impacting on a commercial-scale Sasol-Lurgi FBDB gasifier. *Fuel, 88*(5), 771–779.

24. Uma Rani, R., et al. (2020). Chapter 5 – Thermochemical conversion of food waste for bio-energy generation. In J. R. Banu et al. (Eds.), *Food waste to valuable resources* (pp. 97–118). Academic Press.

25. Higman, C., & Tam, S. (2014). Advances in coal gasification, hydrogenation, and gas treating for the production of chemicals and fuels. *Chemical Reviews, 114*(3), 1673–1708.

26. Murthy, B. N., et al. (2014). Petroleum coke gasification: A review. *The Canadian Journal of Chemical Engineering, 92*(3), 441–468.

27. Amick, P. (2005). ConocoPhillips and gasification's new world–An overview. In *Gasification technologies conference*. Citeseer.

28. Maurstad, O. (2005). An overview of coal based Integrated Gasification Combined Cycle (IGCC) Technology. *Laboratory for energy and the environment* LFEE 2005-002 WP. pp. 2-11.

29. Wang, T. (2017). An overview of IGCC systems. In *Integrated gasification combined cycle (IGCC) technologies*. pp. 1–80.

30. Mills, S. (2012). *Coal-fired CCS demonstration plants, 2012*. IEA Clean Coal Centre.

31. Smith, P. V., et al. (2005). *KBR transport gasifier* (pp. 9–12). Gasification Technologies.

32. Loha, C., et al. (2018). Gasifiers: Types, operational principles, and commercial forms. In S. De et al. (Eds.), *Coal and biomass Gasification: Recent advances and future challenges* (pp. 63–91). Springer.

33. Lee, R. P., et al. (2021). An analysis of waste gasification and its contribution to China's transition towards carbon neutrality and zero waste cities. *Journal of Fuel Chemistry and Technology, 49*(8), 1057–1076.

34. Doe, U. Session 1 gasification-overview-1. *Clean Coal Technology, 1*, 2.

35. Davis, B. H., & Hower, J. (2012). Coal technology for power, liquid fuels, and chemicals. In J. A. Kent (Ed.), *Handbook of industrial chemistry and biotechnology* (pp. 749–805). Springer.

36. Laboratory, National Energy Technology. *British Gas/Lurgi Gasifier.*

37. Ryzhkov, A., et al. (2016). Development of entrained-flow gasification technologies in the Asia-Pacific region. *Thermal Engineering, 63*, 791–801.

38. Guo, M., Song, W., & Buhain, J. (2015). Bioenergy and biofuels: History, status, and perspective. *Renewable and Sustainable Energy Reviews, 42*, 712–725.

39. Asadullah, M. (2014). Barriers of commercial power generation using biomass gasification gas: A review. *Renewable and Sustainable Energy Reviews, 29*, 201–215.

40. Leung, D. Y., Yin, X., & Wu, C. (2004). A review on the development and commercialization of biomass gasification technologies in China. *Renewable and Sustainable Energy Reviews, 8*(6), 565–580.

41. Sansaniwal, S., Rosen, M., & Tyagi, S. (2017). Global challenges in the sustainable development of biomass gasification: An overview. *Renewable and Sustainable Energy Reviews, 80*, 23–43.

42. Pereira, E. G., et al. (2012). Sustainable energy: A review of gasification technologies. *Renewable and Sustainable Energy Reviews, 16*(7), 4753–4762.

43. Sajid, M., et al. (2022). Gasification of municipal solid waste: Progress, challenges, and prospects. *Renewable and Sustainable Energy Reviews, 168*, 112815.

44. Kohli, S., & Ravi, M. (2003). Biomass gasification for rural electrification: Prospects and challenges. *SESI Journal, 13*(1–2), 83–101.

45. Luo, X., et al. (2018). Biomass gasification: An overview of technological barriers and socio-environmental impact. In *Gasification for Low-Grade Feedstock* (pp. 1–15).

46. Kumar, A., Jones, D. D., & Hanna, M. A. (2009). Thermochemical biomass gasification: A review of the current status of the technology. *Energies, 2*(3), 556–581.

47. Mahinpey, N., & Gomez, A. (2016). Review of gasification fundamentals and new findings: Reactors, feedstock, and kinetic studies. *Chemical Engineering Science, 148*, 14–31.

48. Suryawanshi, S., et al. (2023). Parametric study of different biomass feedstocks used for gasification process of gasifier—A literature review. *Biomass Conversion and Biorefinery, 13*(9), 7689–7700.
49. Senthilkumar, V., & Prabhu, C. (2024). Optimization of design and development of a biomass gasifier – A review. *Biofuels, 15*(8), 1–19.
50. Emun, F., et al. (2010). Integrated gasification combined cycle (IGCC) process simulation and optimization. *Computers & Chemical Engineering, 34*(3), 331–338.
51. Felix, C. B., et al. (2022). A comprehensive review of thermogravimetric analysis in lignocellulosic and algal biomass gasification. *Chemical Engineering Journal, 445*, 136730.
52. Rubinsin, N. J., et al. (2024). An overview of the enhanced biomass gasification for hydrogen production. *International Journal of Hydrogen Energy, 49*, 1139–1164.
53. Tezer, Ö., et al. (2022). Biomass gasification for sustainable energy production: A review. *International Journal of Hydrogen Energy, 47*(34), 15419–15433.
54. Giuffrida, A., Romano, M. C., & Lozza, G. (2013). Efficiency enhancement in IGCC power plants with air-blown gasification and hot gas clean-up. *Energy, 53*, 221–229.

Chapter 2
Fundamentals of Gasification Technology

Understanding Gasification: Transforming Matter into Energy by Harnessing Heat, Pressure, and Chemical Reactions

2.1 Introduction

Many decades ago, thermal processes like combustion were widely accepted in many industries. This became more attractive as it became production energy that is highly required for electricity generation. Those approaches are specially aligned to using solid fossil fuels, i.e., coal. After using coal for two decades, it was noticed that a depletion of coal could occur in the near future [1]. This led to finding an alternative source of solid fuel and drove us in the direction of searching for renewable solid fuel sources and making existing technology more sustainable. After a certain period, biomass was found to be an achievable and renewable solid fuel that can be used for combustion and other thermal processes.

In the meantime, while searching for alternative solid fuel from different possible sources, priority was also given to efficiently finding energy and other chemicals, as the previously defined combustion is only desirable for getting energy. Therefore, gasification, scientifically so-called partial oxidation in a limited oxygen environment has emerged. The gasification process shows many advantages over the combustion of solid fuel. Those advantages include getting high-value syngas during gasification, which can be used to generate valuable chemicals [2]. Gasification product gases are less harmful to the environment as toxic gases released are less harmful. Therefore, in the overall briefing of the gasification technology with advantages over other thermochemical conversion processes, the current chapter deals with insights into the fundamental aspects of gasification technology by defining the properties and composition of feedstocks, different products of gasification processes, gasification environments, and classification of different gasifiers.

M. K. Jena, H. B. Vuthaluru, *Gasification Technology*,
https://doi.org/10.1007/978-3-031-71044-5_2

2.2   Properties and Composition of Feedstocks

The properties and composition of solid fuels are some of the primary factors in characterizing different solid fuels, among others. The properties of solid fuel include physical and chemical properties. The physical properties include particle size, surface area, sphericity, and other factors pertinent to physical characteristics. The chemical properties include a wide range ranging from the chemical composition of different constituents, alkali and alkaline earth metal (AAEM) species, and the chemical structure of various components [3, 4]. Based on the advancement of science and technology, in the recent development of new sophisticated equipment, the properties of solid fuel are determined more precisely, enabling us to determine the properties more precisely. Solid fuel composition is also classified as per the proximate analysis, which displays moisture content, volatile matter, fixed carbon, and ash content. In addition, the ultimate analysis was carried out to understand the composition of carbon, hydrogen, oxygen, nitrogen, and sulfur to compare one fuel with another solid fuel. In recent years, other analytical analyses have been employed to find the composition of inorganic content (including the AAEM species content), such as X-ray fluorescence (XRF) spectroscopy (for better vision of solid fuel composition) [5].

2.2.1   Solid Fossil Fuels

Coal is one primary source in the earth's crust and is widely used for gasification and other thermochemical conversion processes. It is formed by plant debris, converted into different forms, and subjected to metamorphic geological changes over time. Many widely accepted theories and concepts exist on the degree of metamorphic change and the coalification process. Those are primarily elaborated on and described in different literature articles and textbooks [6, 7]. In this chapter, an essential elaboration on coal is presented based on the aim of selling fossil fuels. In a burial condition, the initially formed peat coal gets converted into a subsequent form under atmospheric conditions [8]. A resulting step or sequence of coalification [9] is presented in Fig. 2.1.

Also, the elemental composition of change with transformation from one coalification stage to another. The respective elemental compositions for each stage are

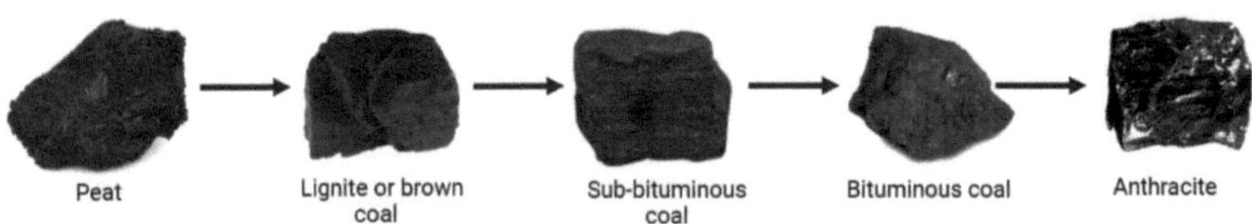

Fig. 2.1 The sequence of coalification with the influence of metaphoric change [9, 10]

Table 2.1 Composition of different types of coal in %w/w dry ash-free basis [11]

Coal type	Carbon	Hydrogen	Oxygen
Peat	58	6	35
Lignite	71	5	23
Sub-bituminous	75	5	16
Bituminous high-volatile	81	6	10
Bituminous low-volatile	88	4	4
Anthracite	94	3	2

tabulated in Table 2.1. It is evident that with coalification, the carbon content in respective coal usually increases adversely, and the hydrogen and hydrogen content decreases. Overall, coal types are broadly classified into high- and low-rank coal depending on the levels of carbon, hydrogen, and oxygen content. The coals that have undergone minimal metaphoric variation in the process of coal formation are categorized as low-rank coal. This indicates that coal with high inorganic ash and low carbon content can be considered low-rank coal. In contrast, coal with high carbon and less ash composition is categorized as high-rank coal.

2.2.2   Renewable Sources (Biomass and Others)

The continuous depletion of solid fossil fuel and possible occurrence in the emission of pollutants through gasification processes lead to finding and using possible renewable sources such as biomass and other sources of high-value carbonaceous material for gasification processes. Renewable sources (which can be stated as renewable solid fuel) include biomass, wood, wood residue, agricultural residue and waste, and organic solid waste from industries and households [12, 13]. The significant advantage of those sources is that they are renewable and less costly to produce than fossil fuels. However, some heterogeneity and contaminants exist due to the diversification of sources. For example, the solid waste generated from households and other subarea like medical and water treatment plants, the composition of feedstock varies greatly with heterogeneity in properties. Some pretreatment methods are adopted before pyrolysis to produce char/biochar and for direct gasification processes. On the other hand, some renewable solid sources have some heterogeneity but lead to less challenge for thermochemical conversion compared to others. The additional pre-treatment method would be time-consuming and create extra operational costs for the primary gasification process [14, 15]. Therefore, selecting appropriate feedstock for the gasification process is essential, and many parameters must be considered before choosing any for gasification processes. In a lab-scale analysis, one can choose any renewable source; however, while considering scaling up the gasification technology/process from a lab scale to a pilot plant scale, some of the considerations need to be discussed, as shown in Fig. 2.2. Although there are certain advantages to selecting a renewable process for gasification, some factors

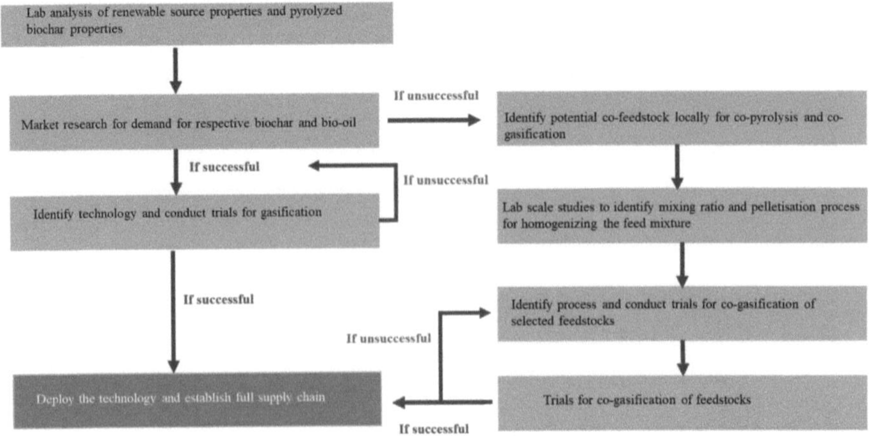

Fig. 2.2 Strategy for selecting the renewable sources for gasification and co-gasification processes

need to be considered. The properties and quantity (tons per year) are the first and primary factors to be considered before executing the gasification and co-gasification processes for the selected feedstock.

2.3 Products of Gasification Technology

Gasification technology includes pyrolysis, the thermochemical conversion of solid fuel into solid fractions called char/biochar, and liquid fraction volatile matter. The subsequent contact of char/biochar and volatile matter with a gasifying agent leads to gasification reactions such as char gasification and volatile reforming reaction in the formation product gas. In emphasizing the overall processes, the pyrolysis (with the evolving of volatile matter) process is very fast compared sub process [14, 16]. Hence, from the kinetic point of view, pyrolysis will not be a rate-limiting process for gasification. Overall, the gasification product gas is the final product, whereas the other components, such as char/biochar and volatile matter, can be considered as a product based on the requirement (Fig. 2.3) [2, 17]. The explanation of each of the products from a gasification process is discussed in the following section of the chapter.

2.4 Pyrolysis

Pyrolysis is converting (in a thermo-chemical conversion manner) a solid fuel into solid carbon char/biochar and volatile matter, which, upon condensation, form bio-oil/tar. The execution of pyrolysis processes is instant and carried out in an inert

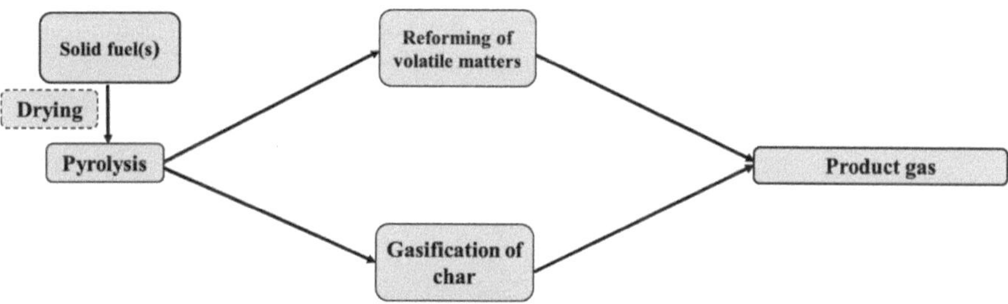

Fig. 2.3 Schematic flow of gasification processes

atmosphere [18, 19]. In general, the pyrolysis operation is usually executed at the beginning of any thermochemical conversion of solid fuel. The condition of the gasification operation determined the operating condition of a pyrolysis process. In some cases, it might be high, low, and equal to the gasification process based on the requirements of the processes. In the inclusion of char and condensed volatile matter, some formation product gas occurs due to the thermal cracking reaction of some volatile matter-pyrolysis of solid fossil [20]. In a normal operation, the pre-treatment method is integrated into the gasification unit before the pyrolysis unit and/or completed gasification unit. In recent advancements in basic needs and demand, some gasification operations are directly carried out in direct of solid fuel with a gasifying agent. In some instances, char gasification is preferred; under that situation, initial pyrolysis was carried out in the inert environment to form char/biochar. Then, it is followed by gasification in contact with a gasifying agent with char to form valuable gas and energy products.

2.4.1 Gasification

Gasification is simple contact of gasifying and solid carbonaceous materials in a thermal environment. The choice of the gasifying environment should be in such a way that it should show limited oxidation (partial oxidation of carbon) capacity in comparison to the combustion process. The process usually produces a gas mixture, partially gasified char/biochar, and residual inorganic content (ash). The composition of products from any gasification process primarily depends on the type of gasifying environment, type of solid fuel, and operating conditions [21]. However, there are some additional benefits to subjecting the hydrodynamics of the reaction environment, which depends on the type of chemical reactor. Upon subjecting to the intrinsic kinetics, the reaction conditions of any particular operating condition depend on the chemical properties of reacting solid fuel and gasifying agent. The following sections explain product composition, the effect of a gasifying environment on product gas composition, and an industrial application of gasification technology for the production of high-value syngas and product fuel for further application [22].

2.4.2 Product Composition Based on Gasifying Feedstock

The products from gasification are product gas, partially gasified char, and solid inorganic residue. The composition of the product gas mixture depends on the gasifying environment and operation conditions, with less dependence on the type of solid fuel. However, the composition of inorganic ash residue and partially gasified char significantly varies with the kind of feedstock. For example, during the gasification of high-ash coal, the residual char matrix at any point of gasification and final ash residue after complete conversion could have a different composition than the gasification of high-rank fuel and/or biomass. Similarly, the gasification of biosolid received from water industries has significant contaminants and heavy metals; hence, in the gasification process, the product gas composition differs from other organic waste gasification [23]. Also, during the thermal treatment of municipal solid waste, the product gas, inorganic ash residue, and others will vary significantly from others due to the heterogeneity and content of different plastic wastes. Further, solid fuels (mostly coal) that are rich in sulfur (S) show the formation of sulfur-containing product gases, which is not desirable in some cases compared to others [24]. In general, it could be stated that in the case of renewable sources of solid fuels, the composition of gasification products (product gas composition, ash residue, and partially gasified char) could vary significantly from each other and from solid fuels in any particular gasifying environment.

2.4.3 Effect of Gasifying Environment on Product

The effect of a gasifying environment on product gas compositions has a more significant impact. Many gasifying environments exist, such as O_2, H_2O (steam), H_2, and CO_2. Those gasifying environments are, in some cases, used individually and in mixed scenarios, mixing one with another. In the case of a gasifying environment such as O_2, the key product gases would be CO and CO_2. In the gasification of solid fuel in a steam environment, the product gases usually contain CO, CO_2, and H_2. The primary gasification reaction determines the formation of key product gas and its change in the gasifying environment. In a steam environment the major reactions are $C(s) + H_2O(g) \rightarrow CO(g) + H_2(g)$, $C(s) + 2H_2O(g) \rightarrow CO_2(g) + 2H_2(g)$ and $CO(g) + H_2O(g) \leftrightarrow CO_2(g) + H_2(g)$. Those steam gasification reactions are subjected to the formation of key product gas-associated dissociation, surface reaction, and desorption of steam molecules. In contrast to that for a solid fuel-O_2 gasification reaction, $2C(s) + O_2(g) \rightarrow 2CO(g)$, $C(s) + O_2(g) \rightarrow CO_2(g)$ and $2CO(g) + O_2(g) \leftrightarrow 2CO_2(g)$ is subjected formation of CO and CO_2 as key product gas [25]. In the case of solid fuel-O_2 heterogeneous reaction, there could be some possibilities for the formation of hydrogen, but not like solid fuel-H_2O heterogeneous reactions. A significant variation in product gas composition is usually observed during gasification in varying gasification environments. In addition, it is also reasonable to state that a variation of key product gas linked with the concentration of the gasifying environment. This means that the formation of product gas gasification of coal char in a steam environment. However, the concentration of CO_2 and H_2 depends on the concentration of steam in the gasifying feed agent. However, a decrease in steam concentrations will likely lead to low hydrogen production with the CO/CO_2 molar flux ratio in the outlet stream significantly affected [26, 27]. This means that in addition to a gasifying agent's partial contribution, the pathway for product gas formation also gets impacted by the product concentrations. The following chapters will explain the reaction mechanisms of varying the path of product gas formation with a change in partial pressures of gasifying agents.

2.5 Gasification Environments

The interaction between solid fuel and gasifying agent at a specified temperature is essential for gasification. Choosing or selecting a gasifying environment for a proper gasification reaction depends on many factors or conditions. To have a gasification reaction carried out at a specified temperature primarily depends on the operating temperature, type of solid fuel, the requirement of key product gas, and the role of the gasification process (individual operational unit or sub-part in integrated processes). In general, the pyrolysis process can be counted as the sub-thermal process of an overall gasification process. Based on that, the environment of thermal processes was treated in two ways: one is an inert atmosphere for pyrolysis, and the other one is an actual gasifying environment if the char/biochar gasification process was the prime intention after pyrolysis. In cases where only gasification was required/or direct gasification of solid fuel was intended, the choice of gasifying environment was taken only. In addition, on subjecting to the prime gasification mechanism, the gasifying environment was sub-categorized into the oxidizing environment, reducing environment, and mixed gasifying environment for gasification. In the case of a reducing and mixed gasifying environment, the partial oxidation and reduction reactions occur simultaneously and/or in a subsequent manner, followed by one another [28].

2.5.1 Gaseous Environment for Pyrolysis

In principle, the pyrolysis operation is carried out to execute the thermochemical conversion processes in an inert atmosphere to form char/bio-char solid residue, condensed bio-oil/tar, and product gases. To achieve a complete pyrolysis process, an adequate inert environment is needed to be established. Hence, the inert environment, like N_2 (Nitrogen) and Ar (Argon), is used. This kind of inert element environment is suitable for all kinds of atmospheres [18]. However, in recent days, inter-environments have been chosen based on the operating temperature. For example, in the co-pyrolysis fluidized bed heat exchanger operation, an integrated gas producer supplies the heat and inter-atmosphere. A gas producer is a unit where solid fuel combustion occurs in the presence of air. That results in the production of CO_2 and other gases present in the air at higher temperatures. That gas was fed to the pyrolyzed chamber in the belief that gases for gas producers are inter in the pyrolysis chamber because the operational temperature is too high for CO_2 and other gases to react with carbon present in solid fuel. This advanced approach has been made to keep low costs and reuse undesired gas streams from the overall process [29].

2.5.2 *Oxidizing Environments for Gasification*

An oxidizing environment is one in which the atmosphere is enriched with O_2. Gasification in an oxidizing environment requires a standardized O_2-rich atmosphere. In this case, the gasification operation is usually carried out at a relatively low temperature compared to a reducing gasifying environment. This is because a gasification reaction between carbon and an oxidizing climate requires less activation energy than a reaction in a carbon and reducing environment, as the gasification operation is based on the rate of carbon consumption and formation of carbon-containing gases. Gasification in the oxidizing environment occurs faster than in the reducing environment [30]. It has the benefit of producing product gas at a higher rate than other gases. The product gas generated during gasification in an oxidizing environment is mostly comprised of CO, CO_2, and other minor gases.

2.5.3 *Reducing Environments for Gasification*

Reducing the environment for gasification comprises steam (H_2O) and H_2. Reducing gasification conditions generally requires high activation energy to execute the gasification reaction. As noticed from the experimental results, the gasification of solid fuel in a steam environment commences at around 700 °C, where steam acts as a reducing environment. In contrast, solid fuel gasification in an oxygen environment usually begins around 400 °C [25, 31]. A section of a gasifying environment based on operation temperature is very vital. The factors controlling the gasification reaction, such as kinetics-controlled, mixed, and diffusion-controlled, are determined mainly by the type of gasifying agent, whether it is either an oxidizing environment or a reducing environment. Representing the actual controlling factor for gasification is essential for understanding the reaction pathway. For a reducing atmosphere, the gasification mechanism is kinetically controlled in temperature ranging from 700 °C–900 °C as the gasification processes commence around ~700 °C [32, 33]. Meanwhile, the specified kinetically controlled regimen for reducing environment acts as a diffusion-controlled regime for the oxidizing environment as the gasification reaction commences at around 400 °C. From an industrial scenario, a gasification operation is usually carried out to achieve a higher gasification rate; in some cases, mixed gasifying environments are generally preferred for executing a gasification reaction.

2.5.4 Mixed Gasifying Environments for Gasification

The gasifying environment for a gasification process was chosen according to the operating temperature and requirement of the gas product depending on the designed and processed conditions. This section presents the discussion based on desired product requirements. In some laboratory-scale scales, the design includes varying the gasifying conditions by mixing them. However, the concentration of each gasifying agent is usually described and determined in keeping with the type of solid fuel used. In the case of oxygen as a gasifying environment, producing CO and CO_2 at a commencement temperature ranging from 350 °C to 500 °C could be usual [33]. However, adding steam to oxygen could enhance the commencement temperature for hydrogen production and diversify the application of production in various fields due to the presence of hydrogen streams. In addition, some features must be implemented while performing gasification experiments in a lab and industrial scale, which will be discussed briefly in the following chapters. The suitability of choosing a mixed gasifying environment is not laying with oxygen and steam; in some conditions, the mixture of hydrogen, steam, and oxygen is usually chosen. Also, for some instances, a mix of CO_2 and steam was taken for gasification at higher temperatures with the aim of CO_2 utilization [34]. Methane, with a mixture of hydrogen and steam, has been considered a gasifying agent for gasifying low-rank fuel [35]. In principle, so far, the considerations that have been made are primarily based on low-ranking fuels, as those fuels show very different properties and conditions for the consumption of carbon in comparison to high-ranking fuels and others. As listed below, some considerations have been taken while selecting a mixed gasifying environment for gasifying any feedstock.

- The mixture of gasifying agents should be considered, keeping in mind the prime objective of gasification temperature and the desired product gas composition.
- The optimization of different gasifying agents in the mixture must not react in between them while feeding to the gasifying reactor.
- The operation temperature for gasification should be chosen so that the reaction mechanism controlling factor overlaps with the other. This would benefit the gasification process by optimizing the product gas composition.
- Some possibilities exist in changing the path of product gas formation in a mixed gasifying environment compared to an individual due to the overlapping of reaction control mechanisms. This behavior is usually expected to optimize the process conditioning and benefit the product gas composition.

2.5.5 Benefits and the Expected Outcomes of Different Gasification Environments

The prime intention of taking a gasifying environment should be a limited oxidation environment for gasification. This results in the formation of more CO than other undesired product gases such as CO_2 and others. Based on fulfilling the prime objective, some additional benefits of gasification are a higher gasification rate and easier carbon consumption during the gasification of any solid fuel. Specifically, for mixed gasifying environments, the expected outcome, as discussed before, would be the formation of the desired product stream condition with high-end application and implications for unnecessary generated streams from industries for gasification. In general, the expected outcomes from a process are elementary and straightforward in following the benefit of gasification in subjecting to process plant industries. A recent approach has been made, such as chemical looping gasification [36, 37], which has certain advantages in segregating from other individuals and manually making mixed gasifying environments. However, efforts are still being made at a commercial level, and they are reported in recent literature [38–40].

2.5.6 Optimization of Concentration for Gasification Environments

The fundamental approach for optimizing process conditions is dependent on the concentration of the gasifying environment. In the case of an individual gasifying environment, the concentration of the gasifying environment is usually optimized to avoid complete oxidization. Specifically, in the gasification of solid fuel with O_2, the gasifying environment is chosen based on the stoichiometric relation between carbon and oxygen in forming CO ($C + 0.5O_2 \rightarrow CO$). As mentioned above, every mole consumed and produced CO requires half more oxygen to surround it. Considering this condition as the prime objective, the design of the gasifying agent mixture should be such that the overall reaction should be allowed to form CO as much as possible. However, in the case of an individual steam gasification environment (a certain amount of steam balanced with an inert gas such as N_2 and Ar), the approach is being made specifically optimizing the concentration of steam inlet in such a way that the steam consumption should be minimal and the reactor should act as a differential reactor [41] during the process of gasification [42]. In principle, the specification and description of gasifying agents are usually considered to obtain true intrinsic kinetics. The design and selection of appropriate concentrations are necessary for moving towards individual gasified agents such as steam and CO_2. The selectiveness of optimizing the concentration and example was demonstrated for steam gasification [42]. Given below are the listed points for optimizing the gasifying agent concentration in the case of steam gasification.

- It should be noted that the amount of steam consumed during gasification should be minimal.
- The consumption of minimal steam during gasification should be constant throughout the reactor, with constant partial pressure. Therefore, the reactor can be regarded as a differential reactor for the consumption of gasifying agents. A differential reactor is a reactor where a small conversion of reactant is induced to form a product to calculate the kinetic parameters for analysis.
- In the optimization process, the actual steam consumed in the whole process is minimal and determined by the amount of feed, char consumption, and water gas shift reaction during gasification.

The gasification condition should be optimized to minimize the amount of steam consumed. To that end, while selecting the concentration of gasifying agent in a mixture, the optimization would need to be followed, for example, for a steam gasification environment alone. The selection of an appropriate gasifying environment would benefit getting the desired product. However, equal importance needs to be given to creating proper contact between the gasifying agent and solid fuel inside the reactor, which depends on the reactor's hydrodynamic behaviors. That led design and classification of specific reactors, as explained in the following section.

2.6 Classification of Gasifiers

2.6.1 Fixed Bed Gasifier

Fixed bed gasifiers are mainly similar to blast furnaces in industrial gasifiers, and a schematic view of the reactor is shown in Fig. 2.4. The fixed bed gasifier usually operates at around 1000 °C [43, 44]. Considering the hydrodynamics of the contact between solid fuel, which generally feeds from the top, and gasifying, which feeds from the bottom, the reactor was segmented into three sub-parts. This type of gasifier is usually easily operated in small and medium-scale power operations. In comparison with other possible reactors (fluidized bed and entrained bed), it is not elementary to maintain the desired temperature in the reaction zone. Based on the reactor's design, the fixed bed gasifier can be further classified into updraft and downdraft gasifiers. The displayed schematic (Fig. 2.4) is an updraft gasifier as the solid fuel is fed from the top, and a gasifying agent is introduced from the bottom. In the case of a downdraft gasifier, solid fuel is introduced from the top, and the fluid gasifying agent is fed from the top side above.

The product gas generated from the gasification processes is usually integrated with steam turbines for power generation in commercial industrial reactors. However, the gas produced by steam turbines in fixed-bed gasifiers is less applicable for synthesis fuel, chemical, and gas turbine applications due to higher hydro carbons, such as aromatic hydrocarbons and tars [45, 46].

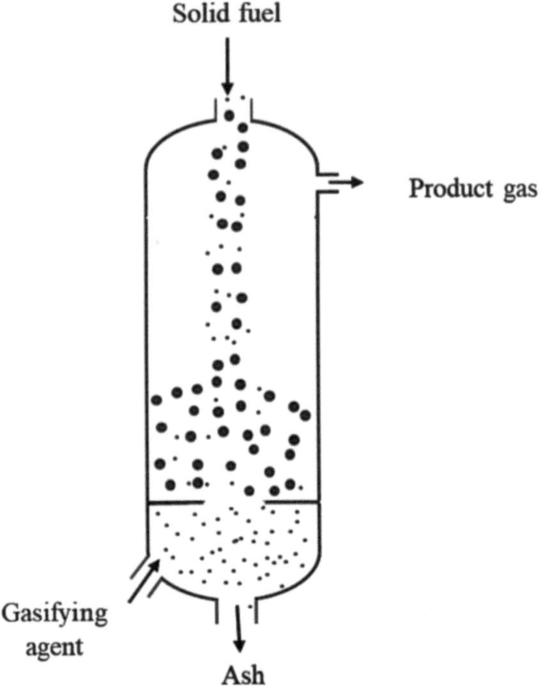

Fig. 2.4 Schematic diagram of fixed bed gasifier

2.6.2 *Fluidized Bed Gasifier*

With specific benefits and extensive use of fixed bed gasifiers, a few challenges have been noted, leading to the development of fluidized bed mixing mechanisms. In general, the fluidized bed gasifier was adopted to overcome the problem associated with previously designed reactors. In a continuous fluidized bed operation, solid feed was fed to the reactor from the top side, and the gasifying agent was fed from the bottom at above minimum fluidization velocity from the bottom. In contrast to fixed bed operation, the fluidized bed operation usually provides intense and vigorous mixing inside the reactor. Also, it offers high contact and residence time between solid fuel and gasifying agents. The evolved or generated product gas is usually removed from the reactor at the top section [47, 48]. The unreacted/inert residual was typically collected from the bottom. A schematic view of the fixed bed gasifier is shown in Fig. 2.5. The prime and key mechanism that makes a significant difference between a fixed bed and a fluidized bed is the hydrodynamics of the reacting bed, mixed ness, and interaction between solid and gaseous reactant. A fluidized bed operation can be continuously operated in a wide range of temperatures, which benefits using any gasifying agent form in kinetics/reaction-controlled regimes.

In addition to the factors mentioned above and the parameters for aging the fluidized bed gasifier, the significant advantage is that a fluidized bed gasifier provides very heat and mass transfer in a bed. High heat and mass transfer are essential for

Fig. 2.5 Schematic diagram of fluidized bed gasifier

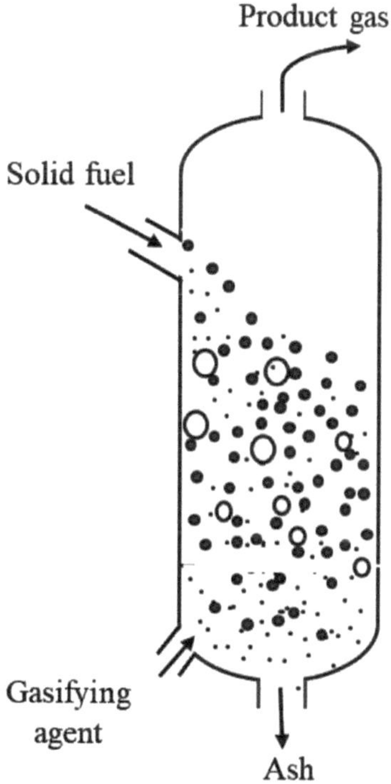

thermal operations such as gasification [49]. There are certain advantages to fluidized bed reactors, such as high heat recovery with efficient product control, utilization of fine solid fuel particles, smaller unit size of the reactor, and better circulation with vigorous mixing of particles in the bed. In another aspect, there are certain disadvantages associated with it. Those include complex designs, making it difficult to scale up from lab cases to commercial plant scale in some scenarios.

2.6.3   Entrained Bed Gasifier

In addressing the problem with a fluidized bed reactor, an entrained flow reactor is designed with diversified application and contact between solid fuel and gasifying agent. An entrained flow gasifier can handle excellent particles in solid phase and/ or slurry, making it a different gasifying reactor. The gasification operation in an entrained bed reactor can be operated from 1000 °C to 1500 °C [50]. That is why entrained flow reactors use specific gasifying agents, which usually limit the reaction at low temperatures, such as H_2, CO_2, and any mixed gasifying agent. In addition, the higher operation of the entrained flow reactor at the higher temperature in the cracking of tar and other higher molecular components was desired for gasification [51]. In contrast, some disadvantages are associated with the entrained bed

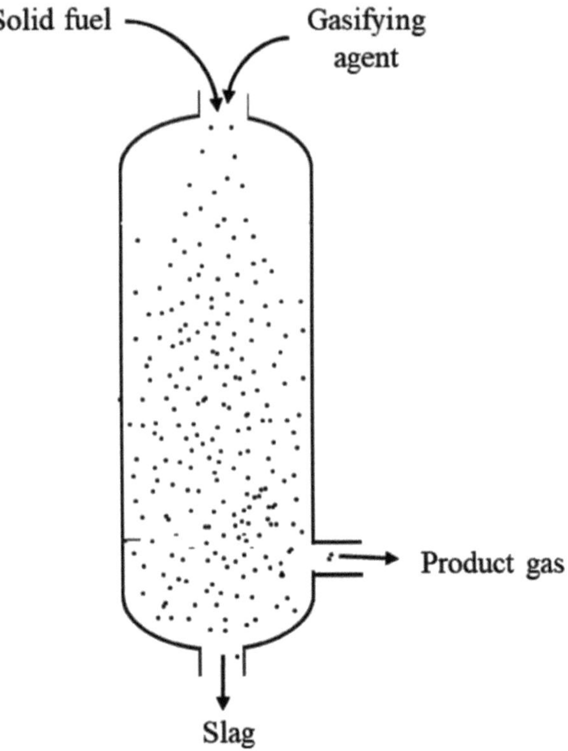

Fig. 2.6 Schematic diagram of entrained bed gasifier

operation to the requirements of pre-treatment processes. As mentioned, the entrained mechanism requires excellent solid fuel, creating cost and limiting application for large-scale plants. A detail of the entrained bed gasifier is schematically shown in Fig. 2.6.

2.6.4 Opportunity and Challenges of Different Gasifiers

The opportunities and challenges associated with different gasifiers broadly differ from each other. Determining opportunities and challenges related to each gasifier are aligned and beneficial to the overall gasification operation in certain aspects [52]. The gasifiers mentioned in the chapter above are classified based on the reaction's design while considering the reacting zone's hydrodynamic and solid-gas interactions between solid fuel and gasifying agent. The opportunities and challenges associated with fixed, fluidized, and entrained beds are limited to hydrodynamic conditions [53]. There is no fixed basis for applying the selective method and reaching a conclusion based on relative comparison. This is because of the diversification in solid fuel properties and their respective interaction with different gasifying agents in varying thermal environments. However, some primary considerations would help evaluate the benefit of a gasification reactor.

- The reactor's scalability should be flexible so that the gasification process aligns with it and can be scaled up from lab to plant scale.
- It should apply to all kinds of feedstock and be able to manage upstream and downstream operations associated with the gasification of any feedstock.
- The capital and operation costs associated with the process during scaling up should be allowed and cost economical.
- The gasifier reactors and processes should be fixable enough to be easily implemented in an integrated process or system.

2.6.5 Progress in the Advancement of Gasifiers

The progress of recent gasifiers has been extensive, and it has been kept as the primary concern in looking at the current demand. Since the population is growing, technology and other resources are increasing simultaneously. A widely accepted feed/solid fuel should be implemented to make the gasification process sustainable. In considering the future potential feedstock, such as e-waste, bio-solids, food organics, garden organics, and other solids apart from biomass and coal [54, 55] the development of gasifiers should be advanced in emphasizing the following concerns.

Gasifying Agent The operation temperature of the gasifier should be widened enough so that all kinds of gasifying agents can be used specific to their kinetics-controlled regimes. For example, in the case of an oxygen-gasifying environment, the gasification commences at around 400 °C, which also leads to the kinetics-controlled regime. At the same time, for the steam environment, the gasification processes majorly commence at around 700 °C with reaction kinetics as the controlled mechanism.

Solid Fuel Particle Size The gasifier and/or gasification reactor can handle various particle sizes. This is because, in certain gasification conditions, bigger particles are desired as particles cannot be reduced very much, such as biomass. In other cases, such as coal, a low particle size is chosen to avoid intraparticle heat and mass limitations.

Hydrodynamics/Mixedness of Particles Inside the Reactor In alignment with chemical reaction engineering, the maximum mix is desirable for solid-gas heterogeneous reactions. However, the reaction kinetics depend on the reaction temperature, and there are some possibilities of differing reaction order with mixed ness factor (early mixing and/or late mixing). The reactor design should be done so that consistency in reaction order should be maintained during gasification for any solid fuel.

Gasification Processes In the recent days of commercialization, the choices made for gasification vary with the process. Direct gasification is usually considered in some scenarios, and char/biochar is generally preferred in some scenarios. Therefore, direct gasification is sometimes stated as one-step gasification, whereas char/biochar gasification is called two-step gasification as char/biochar is subsequently from the pyrolysis process.

The recent gasifier should be made attractive enough to cover all conditions based on prime considerations, such as gasifying agent, solid fuel particle size, and hydrodynamic/ Mixedness of particles inside the reactor and gasification processes. That will further help diversify a wide range of solid fuel/feed and gasifying agents for gasification. This diversification of gasifiers will bring the scalability of the new gasification process to a very smooth commercial level.

2.6.6 Modern Techniques in the Development of Gasifiers for the Future

The current scenario of the development of gasifiers is primarily focused on a sustainable approach. Also, some emphasis was given to the applicability of a wide range of solid fuels and the development of additional processes such that energy can be recovered most economically in addition to product gas yields and partially gasified char/biochar. In some developed countries where gasification was used only for power generation, such as coal power stations, a shift has occurred to using other carbonaceous materials in terms of solid fuel for gasification. In addition, the current approach for gasification is made to overcome the benefit of one reactor over another reactor configuration. In recent efforts, observing the combined reactor, such as a fluidized-fixed bed reactor, would be expected [56–58]. Similarly, other reactors, such as two-stage and three-stage, have been developed where pre-developed biochar was further utilized as catalysts for tar reforming and other gasification reactions. In addition, several different approaches were considered to get additional benefits, which will be elaborated further in Chaps. 4 and 5.

References

1. Horsfield, B. (1979). History of gasification. *Master*.
2. Molino, A., Chianese, S., & Musmarra, D. (2016). Biomass gasification technology: The state of the art overview. *Journal of Energy Chemistry, 25*(1), 10–25.
3. Stewart, B. R., & Daniels, W. L. (1992). Physical and chemical properties of coal refuse from Southwest Virginia. *Journal of Environmental Quality, 21*(4), 635–642.
4. Mastalerz, M., et al. (2009). Effects of coal storage in air on physical and chemical properties of coal and on gas adsorption. *International Journal of Coal Geology, 79*(4), 167–174.

5. Endriss, F., et al. (2023). Analytical methods for the rapid determination of solid biofuel quality. *Chemie Ingenieur Technik, 95*(10), 1503–1525.

6. Bell, D. A., Towler, B. F., & Fan, M. (2010). *Coal gasification and its applications.* William Andrew.

7. Liu, K., Cui, Z., & Fletcher, T. H. (2009). Coal gasification. *Hydrogen and Syngas Production and Purification Technologies,* 156–218.

8. O'Keefe, J. M., et al. (2013). On the fundamental difference between coal rank and coal type. *International Journal of Coal Geology, 118,* 58–87.

9. Diessel, C. F., & Diessel, C. F. (1992). *The coalification process. Coal-bearing depositional systems* (pp. 41–85).

10. Shutterstock, *Coal samples.*

11. Kandiyoti, R., Herod, A. A., & Bartle, K. D. (2006). Chapter 2 – Fossil fuels: Origins and characterization methods. In R. Kandiyoti, A. A. Herod, & K. D. Bartle (Eds.), *Solid fuels and heavy hydrocarbon liquids* (pp. 13–35). Elsevier Science Ltd.

12. Indrawan, N., et al. (2020). Distributed power generation via gasification of biomass and municipal solid waste: A review. *Journal of the Energy Institute, 93*(6), 2293–2313.

13. Yang, Y., et al. (2021). Gasification of refuse-derived fuel from municipal solid waste for energy production: A review. *Environmental Chemistry Letters, 19,* 2127–2140.

14. Pereira, E. G., et al. (2012). Sustainable energy: A review of gasification technologies. *Renewable and Sustainable Energy Reviews, 16*(7), 4753–4762.

15. Sibiya, N. T., et al. (2021). Effect of different pre-treatment methods on gasification properties of grass biomass. *Renewable Energy, 170,* 875–883.

16. Bridgwater, A., & Peacocke, G. (2000). Fast pyrolysis processes for biomass. *Renewable and Sustainable Energy Reviews, 4*(1), 1–73.

17. Li, C.-Z. (2013). Importance of volatile–char interactions during the pyrolysis and gasification of low-rank fuels—A review. *Fuel, 112,* 609–623.

18. Demirbas, A., & Arin, G. (2002). An overview of biomass pyrolysis. *Energy Sources, 24*(5), 471–482.

19. Sekar, M., et al. (2022). Production and utilization of pyrolysis oil from solidplastic wastes: A review on pyrolysis process and influence of reactors design. *Journal of Environmental Management, 302,* 114046.

20. Solomon, P. R., Fletcher, T. H., & Pugmire, R. J. (1993). Progress in coal pyrolysis. *Fuel, 72*(5), 587–597.

21. Breault, R. W. (2010). Gasification processes old and new: A basic review of the major technologies. *Energies, 3,* 216–240. https://doi.org/10.3390/en3020216

22. Higman, C. (2008). Gasification. In *Combustion engineering issues for solid fuel systems* (pp. 423–468). Elsevier.

23. Yu, M. M., et al. (2015). Co-gasification of biosolids with biomass: Thermogravimetric analysis and pilot scale study in a bubbling fluidized bed reactor. *Bioresource Technology, 175,* 51–58.

24. Gryglewicz, G., & Jasieńko, S. (1992). The behaviour of Sulphur forms during pyrolysis of low-rank coal. *Fuel, 71*(11), 1225–1229.

25. Jena, M. K., et al. (2021). Mechanistic insights into the kinetic compensation effects during the gasification of Loy Yang Brown coal char in O2. *Industrial & Engineering Chemistry Research, 60*(49), 17881–17896.

26. Akhtar, M. A., & Li, C.-Z. (2020). Mechanistic insights into the kinetic compensation effects during the gasification of biochar: Effects of the partial pressure of H2O. *Fuel, 263,* 116632.

27. Sidek, F. N., Abdul Samad, N. A. F., & Saleh, S. (2020). Review on effects of gasifying agents, temperature and equivalence ratio in biomass gasification process. *IOP Conference Series: Materials Science and Engineering, 863*(1), 012028.

28. Basu, P., & Kaushal, P. (2024). Chapter 11 – hydrogen production and fuel cells. In P. Basu & P. Kaushal (Eds.), *Biomass gasification, pyrolysis, and Torrefaction* (4th ed., pp. 397–430). Academic Press.

29. Kalpit, S., et al. (2022). A method and system for pyrolysis. *Google Patents.*
30. Baker, R. T. K., & Sherwood, R. D. (1981). Catalytic gasification of graphite by nickel in various gaseous environments. *Journal of Catalysis, 70*(1), 198–214.
31. Jena, M. K., et al. (2021). Studies into the kinetic compensation effects of Loy Yang Brown coal during gasification in a steam environment—A mechanistic view. *Chemical Engineering Journal Advances, 8,* 100159.
32. Akhtar, M. A., Zhang, S., & Li, C.-Z. (2019). Mechanistic insights into the kinetic compensation effects during the gasification of biochar in H2O. *Fuel, 255,* 115839.
33. Akhtar, M. A., et al. (2018). Kinetic compensation effects in the chemical reaction-controlled regime and mass transfer-controlled regime during the gasification of biochar in O2. *Fuel Processing Technology, 181,* 25–32.
34. Li, T., et al. (2017). Effects of char chemical structure and AAEM retention in char during the gasification at 900°C on the changes in low-temperature char-O_2 reactivity for collie sub-bituminous coal. *Fuel, 195,* 253–259.
35. Hiblot, H., et al. (2016). Steam reforming of methane in a synthesis gas from biomass gasification. *International Journal of Hydrogen Energy, 41*(41), 18329–18338.
36. Nguyen, N. M., et al. (2021). Biomass-based chemical looping gasification: Overview and recent developments. *Applied Sciences, 11*(15), 7069.
37. Goel, A., et al. (2022). Biomass chemical looping gasification for high-quality syngas: A critical review and technological outlooks. *Energy Conversion and Management, 268,* 116020.
38. Wang, P., et al. (2015). Chemical-looping combustion and gasification of coals and oxygen carrier development: A brief review. *Energies, 8*(10), 10605–10635.
39. Dai, J., & Whitty, K. J. (2022). Chemical looping gasification and sorption enhanced gasification of biomass: A perspective. *Chemical Engineering and Processing-Process Intensification, 174,* 108902.
40. Shen, X., et al. (2021). Enhanced and environment-friendly chemical looping gasification of crop straw using red mud as a sinter-resistant oxygen carrier. *Waste Management, 121,* 354–364.
41. Ludwig, R., Estévez, L. A., & Smith, J. (1985). The differential-reactor approximation under dynamic conditions. *Chemical Engineering Science, 40*(5), 759–767.
42. Yip, K., et al. (2010). Effect of alkali and alkaline earth metallic species on biochar reactivity and syngas compositions during steam gasification. *Energy & Fuels, 24*(1), 173–181.
43. Harris, D., & Roberts, D. G. (2023). 19 – Coal gasification and conversion. In D. Osborne (Ed.), *The coal handbook* (2nd ed., pp. 665–691). Woodhead Publishing.
44. Warnecke, R. (2000). Gasification of biomass: Comparison of fixed bed and fluidized bed gasifier. *Biomass and Bioenergy, 18*(6), 489–497.
45. Letcher, T., (2022). Comprehensive renewable energy.
46. Luque, R., & Clark, J. (2010). *Handbook of biofuels production: Processes and technologies.* Elsevier.
47. Ram, M., & Mondal, M. K. (2022). Chapter 13 – Biomass gasification: a step toward cleaner fuel and chemicals. In B. Gurunathan, R. Sahadevan, & Z. A. Zakaria (Eds.), *Biofuels and bioenergy* (pp. 253–276). Elsevier.
48. Pfeifer, C., Koppatz, S., & Hofbauer, H. (2011). Steam gasification of various feedstocks at a dual fluidised bed gasifier: Impacts of operation conditions and bed materials. *Biomass Conversion and Biorefinery, 1*(1), 39–53.
49. Fryda, L. E., Panopoulos, K. D., & Kakaras, E. (2008). Agglomeration in fluidised bed gasification of biomass. *Powder Technology, 181*(3), 307–320.
50. Pang, S. (2016). Chapter 9 – Fuel flexible gas production: Biomass, coal and bio-solid wastes. In J. Oakey (Ed.), *Fuel flexible energy generation* (pp. 241–269). Woodhead Publishing.
51. Krishnamoorthy, V., & Pisupati, S. V. (2015). A critical review of mineral matter related issues during gasification of coal in fixed, fluidized, and entrained flow gasifiers. *Energies, 8,* 10430–10463. https://doi.org/10.3390/en80910430

52. Sansaniwal, S. K., Rosen, M. A., & Tyagi, S. K. (2017). Global challenges in the sustainable development of biomass gasification: An overview. *Renewable and Sustainable Energy Reviews, 80*, 23–43.

53. Li, C.-Z. (2007). Some recent advances in the understanding of the pyrolysis and gasification behaviour of Victorian brown coal. *Fuel, 86*(12–13), 1664–1683.

54. Hameed, Z., et al. (2021). Gasification of municipal solid waste blends with biomass for energy production and resources recovery: Current status, hybrid technologies and innovative prospects. *Renewable and Sustainable Energy Reviews, 136*, 110375.

55. Shahabuddin, M., & Alam, T. (2022). Gasification of solid fuels (coal, biomass and MSW): Overview, challenges and mitigation strategies. *Energies, 15*(12), 4444.

56. Singh, S., et al. (2022). An integrated two-step process of reforming and adsorption using biochar for enhanced tar removal in syngas cleaning. *Fuel, 307*, 121935.

57. Singh, S., et al. (2023). Enhanced tar removal in syngas cleaning through integrated steam catalytic tar reforming and adsorption using biochar. *Fuel, 331*, 125912.

58. Jena, M. K. (2022). *Insights into the kinetic compensation effects and kinetic parameters during gasification of low-rank coal char*. Curtin University.

Chapter 3
Engineering Insights into Gasification Technology

Uncovering the Gasification Science and Innovation That Transform Ideas into Reality

3.1 Introduction

Gasification technology is one of the most considerable technologies for the thermochemical conversion of solid carbon material into valuable products, gas, and energy. The gasification process has been well-known to the scientific and industrial community for many decades. It has also been used in many industrial applications as a primary source and a subpart of many integral processes. Based on the scientific or industrial approach, the understanding of the gasification process is still in process. However, the advanced way of interpreting and developing assessment skills for new technologies would have enabled the desired knowledge to be partly fulfilled. As science and technology progress daily, the way we understand processes in the past few decades, in recent years, and at present varies greatly. The current chapter elaborates on an engineering insight into gasification technology. This chapter also briefly explains different gasification segments, enabling us to build knowledge on a gasification process and forecast and determine the future aspects of gasification technology. Here, an idea about gasification reaction and kinetics, mechanism of reaction kinetics in catalytic gasification, gasification thermodynamics, transport phenomena in gasifying bed, and advances in research progress of gasification technology will be discussed.

3.2 Gasification Reactions and Kinetics

Understanding gasification reactions and kinetics is pivotal in enhancing the overall efficiency of a process, be it in an individual or integrated system. The gasification reaction, a crucial determinant, distinguishes whether the process is an actual gasification or deviating from other thermal conversion processes like combustion. The

© The Author(s), under exclusive license to Springer Nature Switzerland AG 2024

M. K. Jena, H. B. Vuthaluru, *Gasification Technology*,

https://doi.org/10.1007/978-3-031-71044-5_3

kinetics, on the other hand, dictate the rate of chemical reaction during the gasification process. Given that gasification reactions occur under varying gasifying environments and temperatures, a comprehensive understanding of these factors is instrumental in predicting performance, designing gasification reactors, and scaling up gasification technologies to an industrial scale.

3.2.1 Solid-Gas Heterogeneous Reactions

The solid-gas heterogeneous reactions primarily occur during solid catalytic reactions or in a catalytic process. In the current section, the significant benefit of subjecting solid-gas heterogeneous reactions is presenting solid fuel-gas (gasifying agent) heterogeneous reactions. In solid-gas heterogeneous reactions, the product gas is not only delivered from solid-gas heterogeneous reactions. A part of the product could get delivered from a gas-gas homogenous reaction as a reaction as there was a possibility of contact between different product gas and gasifying agents under high temperatures. The solid-gas heterogeneous reaction is identified with different names for reactions based on the occurrence of the reaction mechanism. The significant solid-gas heterogeneous reactions are listed below, defined as the basis of the elementary mechanism (Table 3.1).

The gasification studies are performed in different gasifying environments. Solid-gas reactions are most likely to occur in solid carbon and gasifying environments. However, in some instances, such as solid-gas heterogeneous water-gas-shift reactions, the formed CO on the carbon surface participates further in other

Table 3.1 Solid-gas heterogeneous reaction occurred during gasification in different gasifying environments [4–6]

Oxygen environment	
$C + \frac{1}{2} O_2 \rightarrow CO_2$	Carbon partial oxidation
$C + O_2 \rightarrow CO_2$	Carbon oxidation
$C_nH_m + n/2\, O_2 \rightarrow nCO + m/2\, H_2$	C_nH_m partial oxidation
Steam environment	
$C + H_2O \rightarrow CO + H_2$	Water-gas reaction
$CO(s) + H_2O \rightarrow CO_2 + H_2$	Solid-gas heterogeneous water gas shift reaction
$C_nH_m + nH_2O \rightarrow CO + (n + m/2)\, H_2$	Steam reforming
Hydrogen environment	
$C + 2H_2 \rightarrow CH_4$	Hydrogasification
Carbon dioxide environment	
$C + CO_2 \rightarrow 2CO$	Boudouard reaction
$C_nH_m + nCO_2 \rightarrow 2nCO + m/2\, H_2$	Dry reforming
Decomposition reaction of tars and hydrocarbons	
$pC_xH_y \rightarrow qC_nH_m + rH_2$	Dehydrogenation
$C_nH_m \rightarrow nC + m/2H_2$	Carbonization

gasification reactions before desorption. The recent finding through kinetic compensation effects obtained this kind of advanced interpretation [1]. In the current section, the presentation of solid-gas heterogeneous reactions is based on individual gasifying environments; however, it could be a mixed gasifying environment under some situations as per the requirement of gasifying conditions. Furthermore, the gasification operation will most likely occur in mixed gasifying environments in an industrial-scale scenario. From an engineering and scientific view, understanding the limitations of different gasifying environments on achieving a complete gasification operation (desired to produce less CO_2 and other harmful gases) in a broad range of temperature operations is essential. It is more well-known that different gasifying agents have different reaction control factors in any particular temperature range of operation. For example, coal gasification in an oxygen environment mainly commences around 350 °C to 400 °C, whereas steam gasification in the same coal will occur around 700 °C [2, 3]. Hence, the possibility of a solid-gas heterogeneous reaction occurring in a mixed environment containing O_2 and H_2O would be different at 700 °C compared to a gasification reaction in an individual one. This led to a more devastating nature of individual gasification mechanisms and the path of solid-gas heterogeneous reactions. Apparently, under a mixed gasifying environment system, a solid-gas heterogeneous reaction ultimately outweighs some possible occurrences of a homogenous reaction—that may lead to encountering changes in the paths of solid-gas heterogeneous reactions. As can be explained for the coal-H_2O solid gas, the heterogeneous reaction is controlled kinetically, whereas the coal-O_2 reaction controls the diffusion mechanism at a temperature of around 700 °C. Therefore, the coal-O_2 reaction is likely to occur faster compared to coal-H_2O and result in navigation to combustion reactions.

Understanding significant gasification reactions is very important as it can allow the step to progress. Although it would have been essential to retrospect the gasification process through elementary reactions in an actual way, some non-elementary reactions always occur at the micro-kinetic level. A study on the kinetic compensation effect and the variation of apparent kinetic parameters with conversion would partly tell us how radicals form during a solid-gas heterogeneous reaction. A thorough understanding of the kinetic compensation effect and the benefit of studying the mechanistic view of gasification reaction will be discussed in the upcoming section of this chapter.

3.2.2 Gas-Gas Homogenous Reactions

A representation of the gas-gas homogeneous reaction during the occurrence of the gasification mechanism in the different gasifying environments is shown in Table 3.2. Mechanistically, an expectation of a gas-gas homogenous reaction is undesirable as it mostly encounters the formation of CO_2 in the case of solid-O_2 reactions. However, in some instances, it may be desirable to have a gas phase homogeneous reaction, such as in the case of the occurrence of a water-gas shift

Table 3.2 Gas-gas homogeneous reaction occurred during gasification in different gasifying environments [6, 7]

Oxygen environment	
$H_2 + \frac{1}{2} O_2 \rightarrow H_2O$	Hydrogen oxidation
Steam environment	
$CO + H_2O \rightarrow CO_2 + H_2$	Water gas shift reaction
$CH_4 + H_2O \rightarrow CO + 3H_2$	Steam methane reforming reaction
Hydrogen environment	
$CO + 3H_2 \rightarrow CH_4 + H_2O$	Methanation

reaction; the occurrence of this leads to forming more H_2 in a steam gasification environment. However, similarly, this may lead to more CO_2 forming in the product gas composition. Hence, approaches should be given to optimize the process parameter, design the reaction to have the required contact/residence time between the gaseous molecules, and ensure its suitability for mixing different gasifying agents. The suitable mixing of two gasifying agents and optimization of any particular gasifying agent, i.e., H_2O, to the desired concentration, need lab-scale experimentation. For example, the gasification of biomass in a steam environment of 15%H_2O-Ar (balanced with argon) usually shows a high gasification rate compared to gasification at 8%H_2O-Ar. Contrary to that, higher rate gasification at 15%H_2O-Ar usually displays less CO/CO_2 molar flux ratio than gasification at 8%H_2O-Ar. This indicates a proper understanding of the path of the product gas in a particular gasifying environment is not always the same. It may change with gasifying agent concentration, temperature, and variation in reactor design.

3.2.3   Role of Gasifying Agent Partial Pressure on Yield and Kinetics

The basis for defining the rate law of any reaction, for example, the solid-gas heterogeneous reaction in the current study scenario, is mainly characterized by the function of temperature, concentration (liquid), and pressure (gas). In a gasification reaction, the gasifying agent is the limit reactant between the reactants, i.e., solid fuel and gasifying agent. A chemical engineering study states that the concentration of limiting reactant should be less (or equal) than the value of the true stoichiometric limiting condition relative to excess reactant, solid fuel, in gasification operation. The concentration of limiting reactants always controls the rate and mechanism of chemical reactions. In a scenario, a gasifying agent primarily drives the gasification operation. Therefore, the partial pressure of the gasifying agent is one of the prime concerns to be considered, along with other factors that optimize yield and reaction kinetics. As seen in Eq. 3.1, $g(p)$ is a function of partial pressure, which plays a substantial role in defining the reaction rate (r). Where $f(x)$, the conversion dependent factor depends on all the properties of solid fuel that change during the gasification reaction.

$$r = f\left(\varkappa\right)g\left(p\right)A\exp\left(\frac{-E}{RT}\right) \tag{3.1}$$

In further elaborating on the impact of gasifying agent partial pressure on yield and reaction, it can be stated that much work has been done in the fundamental investigation of reaction mechanisms [8, 9]. The steam gasification performed for biomass particles (Figs. 3.1 and 3.2) at two different gasifying environments of low and high partial pressure steam suggests that low partial pressure steam leads to a low char gasification rate and evenly affects the formation of CO, CO_2, and H_2. This indicates that higher concentrations of gasifying agents allow it to be adsorbed on multiple active site surfaces to become more compared to lower concentrations of gasifying environments (Fig. 3.1).

In contradiction, the higher consumption rate of carbon active in biomass biochar may lead to a change in path product gas formation and acting in the domination of one reaction over another—that subsequently results in a change in relative product gas composition and yield of desired product gases. As seen in the lab-scale experimental analysis, the molar flux ratio of CO/CO_2 varies with changes in the partial pressure of the gasifying agent (Fig. 3.2 a–d). At any particular condition (temperature), the higher rate of biomass char gasification at higher steam partial pressure ($15\%H_2O$-Ar) resulted in a low CO/CO_2 molar flux ratio in final product gas composition compared to gasification at $2\%H_2O$-Ar. This phenomenon occurring in steam gasifying environments may/may not happen in other cases. However, it can be advisable that a higher rate of gasification is not always recommendable for industrial-scale operations. Based on the prime objective of gasification operation, it can be suggested that a fundamental investigation to identify the impact of gasifying agent partial pressure on the rate of a chemical reaction and yield of the desired product is essential. It can be questioned what condition will be where there is no impact of gasifying agent partial pressure on rate gasification. This will be further discussed in the following sections of the current chapter.

3.2.4 *Effect of Temperature on the Yield and Kinetics*

The correlation of temperature with gasification yield and kinetics is not linear. The gasification kinetics and rate are not directly approachable to the temperature [10]. There would be some interdependency between the kinetics and temperature, but approximation and interpretation of yield and mechanism require further understanding. Temperature and pressure are the primary factors to commence a chemical reaction. Let us analyze a condition where pressure is constant, and the gasification rate is driven by temperature variation. In each gasifying environment, the gasification reaction commenced at a particular temperature; for O_2, the gasification reaction commenced at ~400 °C. Meanwhile, for H_2O (steam) and CO_2, the gasification reaction usually commenced at 700 °C and 1000 °C, respectively. The difference in

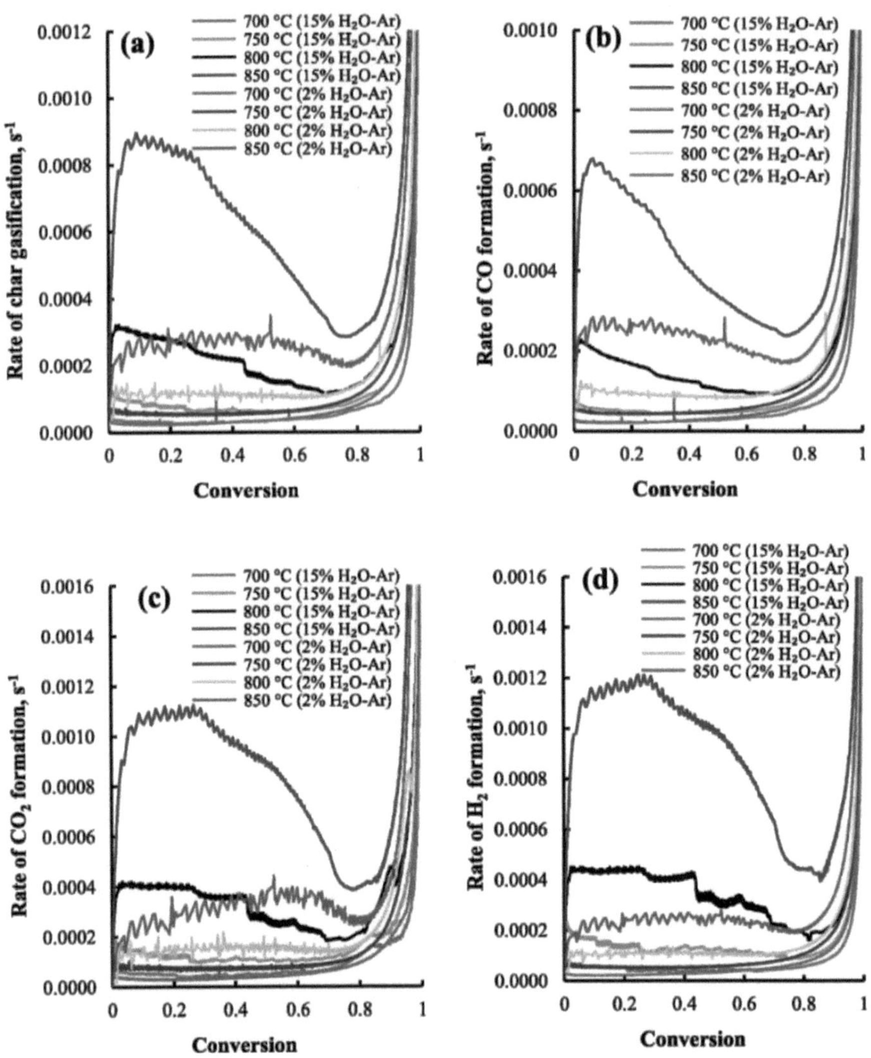

Fig. 3.1 a–d Rate of char, CO, CO_2, and H_2 vs conversion at steam environment of 15%H_2O-Ar and 2%H_2O-Ar (balanced argon) [9]

gasification reaction commencement temperature shows a different temperature range for their respective kinetics, mixed and diffusion-controlled mechanisms. Three regimes are defined per lab-scale experimental investigation: kinetics-controlled, mixed, and diffusion-controlled. The kinetics-controlled regime is a low-temperature regime with respect to that particular gasifying agent and is dominated/controlled by chemical characteristics/properties of solid fuel—that changed chemical characteristics and properties vary/change linearly with temperature. The figure below shows a representativeness of the sharp slope of the Arrhenius plot (Fig. 3.3a–b).

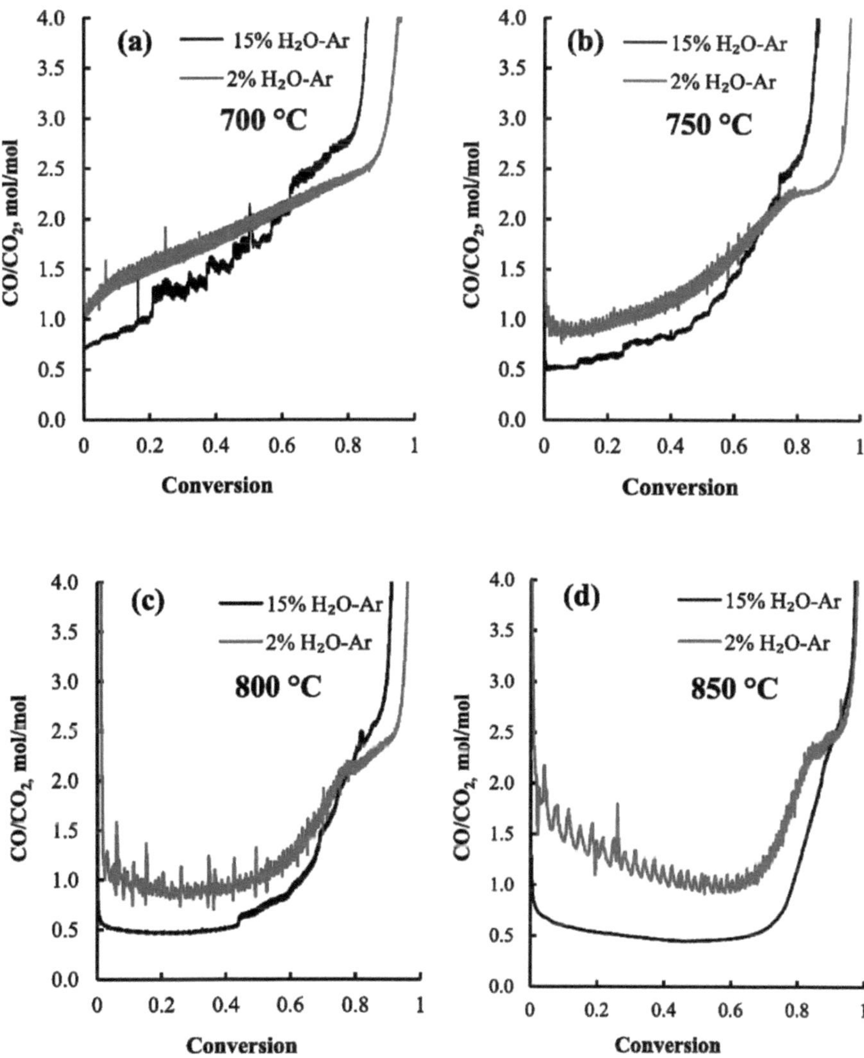

Fig. 3.2 a–d Molar flux of CO/CO$_2$ at 700 °C, 750 °C, 800 °C, and 850 °C during biomass char gasification at 15%H$_2$O-Ar and 2%H$_2$O-Ar [9]

Moreover, proceeding further beyond the kinetics-controlled regime, a feasibility of diffusion-controlled reaction can occur. In the mixed regime, the solid-gas gasification mechanism is controlled by both mass transfer limitation and chemical properties of solid fuel. Although there would have been an increase in the gasification reaction rate, the proportional change in rate with an increase in temperature is not as significant as in the kinetically controlled regime. This comes from the dominance of mass transfer limitation (which was signified by the physical properties of solid fuel) on chemical kinetics of the gasification reaction—which leads to the

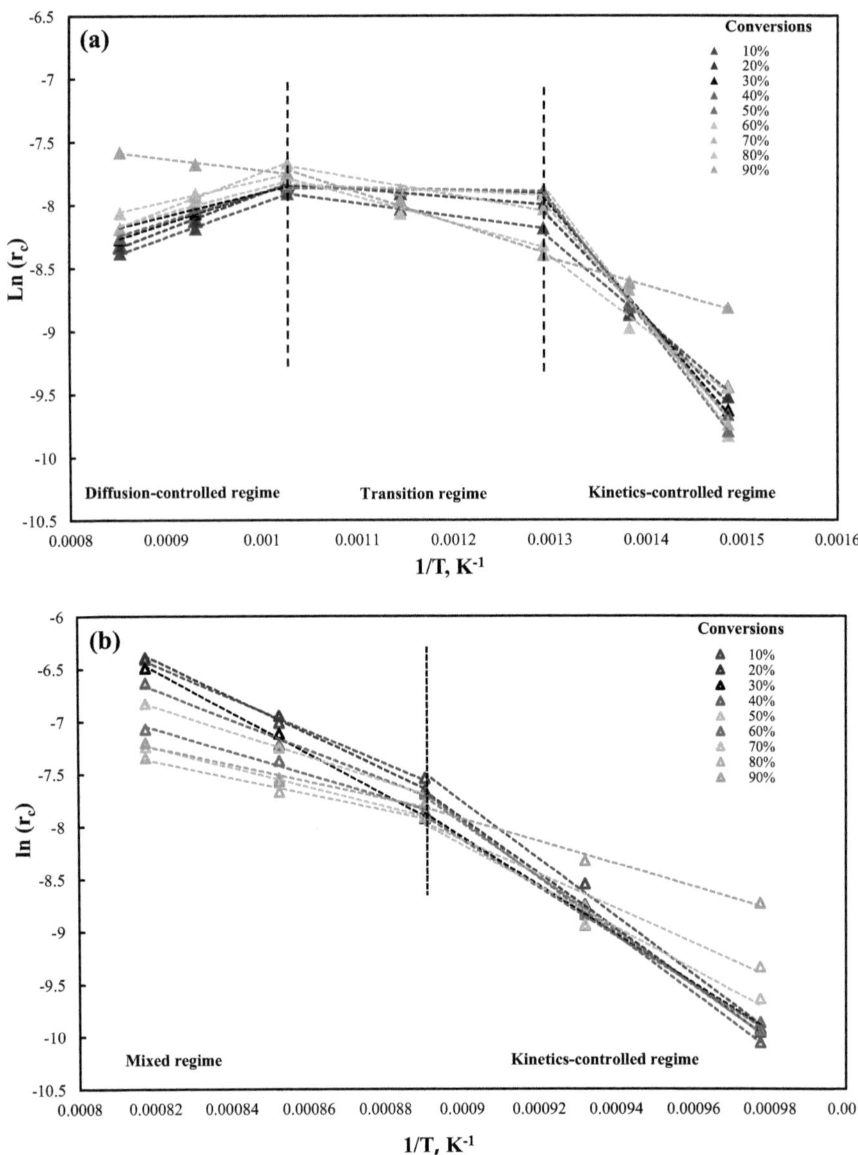

Fig. 3.3 Arrhenius plot of coal char gasification in an O_2 environment (**a**) and an H_2O environment (**b**) [1, 2]

generation of non-intrinsic kinetics throughout the gasification reaction. Like the kinetic-controlled regime, this occurrence of mixed regime varies with the gasifying agent for a particular kind of feedstock, as shown in Fig. 3.3a–b. After going further, a diffusion-controlled mechanism is likely to occur where a steep, non-significant change rate of gasification reaction with temperature is noticed. In the case of the char-O_2 gasification reaction (Fig. 3.3a), a diffusion-controlled regime was seen from 600 °C to 900 °C. The physical properties of solid fuel particles purely control a diffusion-controlled regime mechanism. In the case of diffusion mechanisms, though the gasification kinetics are not desirable, some industrial gasifiers are willing to operate because of the higher gasification reaction rate.

Gasification kinetics are not always influential in chemistry, which we need to look for. In the industrial gasifier, the commercial viability of a thermal technology is primarily dependent on the prime objective, and yield is one of them. A lab-scale study performed low-rank coal char of two different particle sizes and compared it with demineralized coal (H-form coal).

The study was performed in char-O_2 gasification reaction, and attention was given to co-relating the yield of key products with factors affecting the yield of gasification kinetics. As we know in an intrinsic way, in a kinetics-controlled mechanism, including the change in chemical properties of char, the catalytic effect of inherent alkali and alkaline earth metal species (AAEM) species plays a primary role in enhancing the gasification rate and formation of key product gas. However, it can be questionable whether the increase in gasification rate will favor the desired product gas formation. It could be yes or no, based on how the path product gas formation was affected by an increase in the reaction rate. As seen from Fig. 3.4a–c, in the case of H-form coal, the rate of char gasification is usually slow because of less catalytic effect AAEM species. The increase of particle size as more significant chars results in higher retention of AAEM species. However, from another aspect, it can be seen that the product CO/CO_2 molar flux ratio becomes lower with increased particle size. The high molar flux ratio is usually obtained for demineralized coal

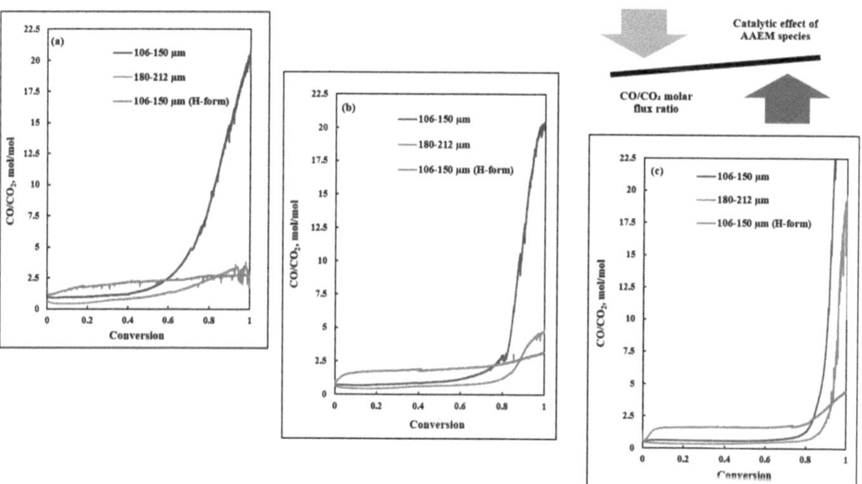

Fig. 3.4 Effect of intraparticle diffusion limitation on CO/CO_2 ratio in the kinetics-controlled regime at 400 °C (**a**), 450 °C (**b**), and 500 °C (**c**) [2]

with a fixed particle size. This advises us that the gasification kinetics with the desired path of product gas formation is essential to achieve the desired yield. A fundamental mechanism understanding the gasification kinetics is critical for optimizing the gasification reaction.

3.2.5 Properties of Char on Gasification Mechanisms

Char gasification is a controlling step in the overall process of prioritizing the overall gasification reaction. Enhancing the efficiency of char gasification reaction means improving the efficiency of overall gasification processes. The functionalized properties, including the physical and chemical properties, led to continuous change during the gasification reaction. Those continuous changes in char properties further affect the gasification reaction kinetics and could cause an alteration of product gas formation. Several chemical and physical properties change during char gasification. Those properties include surface functional groups, carbon bonds, carbon structure, available surface area, pore size, and pore volume [11, 12]. As can be seen from Fig. 3.5, there is a continuous change in char structure (Fig. 3.5a) with a change in yield during the gasification of char in a steam environment. This change in total Raman area becomes low and changes with variation in gasification temperature. Those speculated changes in char structure here in the case of char-H_2O reaction might not likely happen in the case of another gasifying environment. However, surprisingly, the trend of change in properties with change in gasification temperature is nearly the same (Fig. 3.5a). This might indicate that the experimented temperature lies in the single controlled mechanism. Moreover, a significant change in char chemical properties is being noticed in case of change in the gasifying environment, i.e., pure, 15%H_2O-Ar (balanced with Ar), and 15%H_2O-CO_2 (Fig. 3.5b) at 800 °C. Specifically, the char-CO_2 reaction shows less change in char structure than the other two conditions, suggesting that the steam gasification reaction affects the char structure more than CO_2 gasification. This must be due to the very low rate of char-CO_2 reaction as it commenced at a very high temperature, i.e., > 800 °C.

Moreover, the changes speculated in char properties also impact AAEM species' retention. Figure 3.5c depicts the rise in Na retention with a decrease in char yield. This indicates that with the increase in char conversion during the process of gasification, the continuous consumption of carbon and changes in char properties led to an increase in the retention of AAEM species inside the char matrix. The higher gasification rate at higher conversions, observed in most gasification processes, might be possible. This AAEM retention mechanism is valid for all kinds of gasification environments. Also, it can be suggested that a higher catalytic effect of AAEM species on gasification reaction might cause the observance of low activation energy at a higher conversion level. However, this stated suggestion might not always be valid as the catalytic effect of AAEM also depends on its chemical form in the char matrix, which will be explained thoroughly in the following section of this chapter.

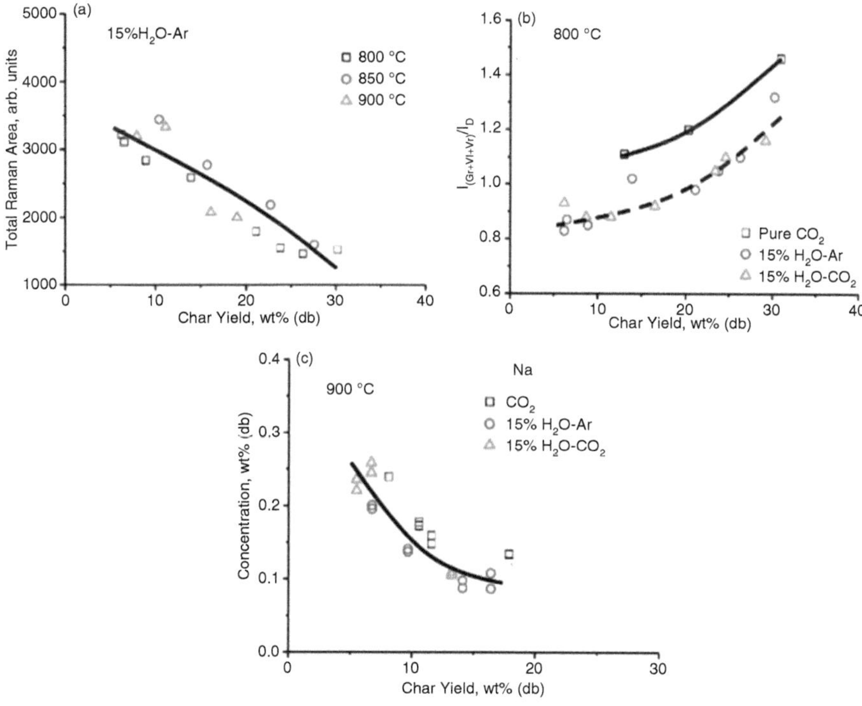

Fig. 3.5 The change in char structure at different yields with change in temperature for a fixed gasifying environment (**a**) with change in gasifying environment for a fixed temperature (**b**) and change in AAEM species (Na) with yield at different gasifying environments (**c**) [13]

3.2.6 Insights into Gasification Mechanisms with the Controlled Regimes

A controlled mechanism primarily drives the overall gasification process. As discussed in our previous sections, there are three controlled mechanisms for gasification reactions. Those are kinetics-controlled, mixed, and diffusion-controlled regimes. The kinetics-controlled regime is driven by the chemical reactivity of char, which is determined by inherent and continuous changes in the chemical properties of char. In a mixed regime, kinetics-controlled parameters dominate the reaction mechanism and mass-transferred controlled parameters. Both properties dominate this mean: the chemical properties, which are contributed by kinetically controlled, and physical properties, which are contributed by diffusion-controlled mechanisms. Finally, in a diffusion-controlled regime, the reactions are controlled by the physical properties of gasifying particle/char/biochar. As mentioned, in a kinetics-controlled regime, the change in chemical properties are the properties that can enhance the formation of the active sites on char surfaces in a continuous consumption of carbons during the execution of solid heterogeneous gasification reactions. For example, the formation of O-containing functional on the char surface, acting as an active for the adsorption of the gasifying agent on it, and acting as a C(O) activated complex for further participation in gasification reaction with other adsorbed species on the surface of the. In addition, the chemical structure of the char matrix, such as the distribution of the carbon chain and aromatic ring, is also essential. The aromatic ring structure and connected carbon chain specify the feasible reactiveness of char in that particular gasifying condition. Based on that, the possible active site formation, carbon consumption, and char matrix rupture can be predicted.

In a similar phenomenon, the physical properties such as surface area, pore size, pore volume, and others are determined to be the primary factors in controlling the mass transfer limitation mechanism in a mixed and diffusion-controlled regime. However, the controlling ness factor on an overall gasification mechanism usually does not directly come from gasifying char's chemical and physical properties. A few other factors drive the gasification reaction to the situation in favor of kinetics-controlled, mixed, and diffusion-controlled behavior. Those factors include temperature, particle size, the turbulence or velocity of the gasifying agent surrounding the particle, and another secondary external factor. The presented kinetics are studied in different regimes for biomass gasification (varying particle size), ranging widely from kinetics-controlled to diffusion-controlled regimes, including the mixed regime (Fig. 3.6).

The study of biomass biochar-O_2 reaction indicates that the low apparent activation energy and apparent frequency factor for bigger biomass particle size, i.e., 2.0–3.35 mm, compared to smaller ones, i.e., 0.8–1.0 mm in kinetics-controlled regime (400 °C, 450 °C, and 500 °C) (Fig. 3.6a–b). The low kinetic value for bigger particle sizes indicates the presence of intraparticle diffusion limitation. In the case

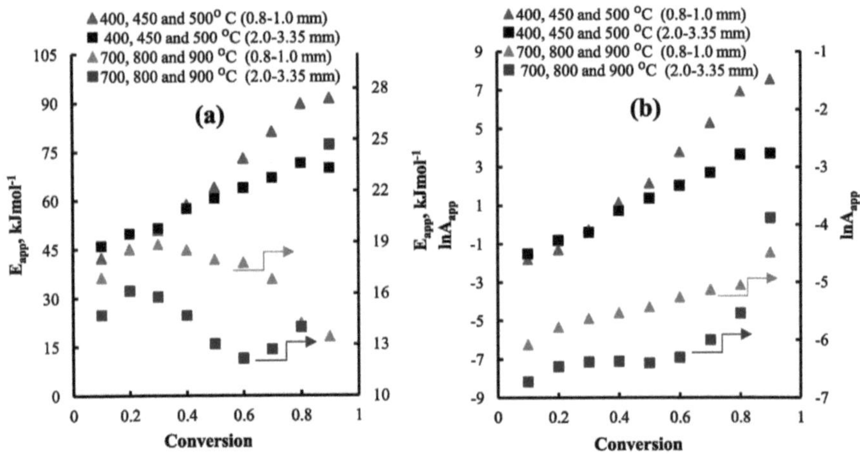

Fig. 3.6 The apparent activation energy (E_{app}) as a function of char conversion over different temperature ranges for 0.80–1.0 mm (**a**) and 2.0–3.35 mm particle sizes and (**b**) The apparent pre-exponential factor ($\ln A_{app}$) as a function of char conversion over different temperature [10]

of smaller particles, there are no intraparticle diffusion limitations. Moreover, the intraparticle diffusion limitations in bigger particles are attributed to the diffusion barrier for gasifying agents and product gases to pass inside the pore of the char matrix. This internal and intraparticle diffusion barrier mechanism is usually noticed at low-temperature regimes because the reaction is kinetically controlled with no external mass transfer limitations, and the substantial effect of change in kinetic parameter with particle size is attributed to the internal diffusion barrier. The presence of intraparticle diffusion limitation in attributing the variation of kinetic parameters to changes in particle size also impacts the path of product gas formation. In addition, intraparticle diffusion limitation may or may not be observed in other conditions for a particular kind of solid fuel particle in a fixed gasifying environment. This means that the presented biomass particle, in interaction with gasifying agent O_2, shows the intraparticle diffusion limitation. However, this intraparticle interparticle diffusion limitation was not observed while analyzing the experimental study for the same particle (0.80–1.0 mm and 2.0–3.35 m) in a steam environment (Fig. 3.7a–b).

On substantially affecting the value of reaction individual kinetics of carbon consumption and product gas formations. As shown in Fig. 3.7a–b, though there is a continuous decrease in kinetics parameters with conversion, no significant difference in kinetic parameters was noticed with a change in particle size. This confirms the absence of intraparticle diffusion limitation in the case of biomass particles during biomass char gasification in a steam environment. Also, it suggests that intraparticle diffusion limitation depends on the gasifying environment and gasification condition, not particle size.

Moreover, as the study has been done intensively for kinetics-controlled regimes, some investigations for mixed and diffusion-controlled regimes are desirable as most industrial reactors operate a diffusion controlled/higher temperature regime to achieve a higher gasification rate. In a diffusion-controlled regime, the rate is

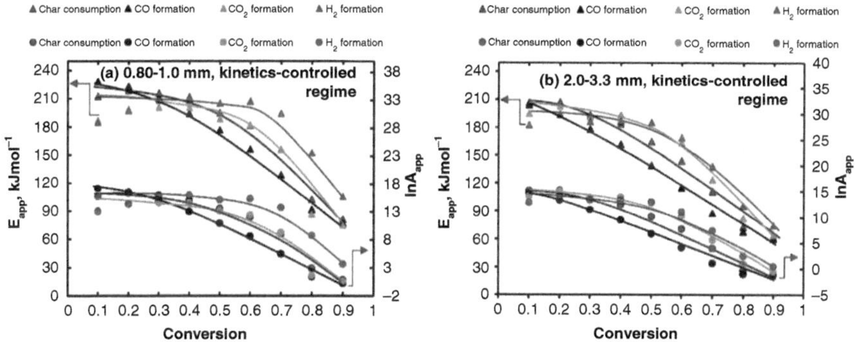

Fig. 3.7 Apparent activation energy (E_{app}) and the apparent pre-exponential factor (lnA_{app}) of bio char consumption and formation of CO, CO_2, and H_2 as a function of biochar conversion in the kinetics-controlled regime for particle sizes of (**a**) 0.80–1.0 mm and (**b**) 2.0–3.3 mm for the biomass biochar-H_2O gasification reaction [14]

controlled by chemical and physical properties; hence, finding true intrinsic kinetic for understanding the reaction mechanism is very complex. However, the recent advancement of analytic techniques and ways of interpreting experimental data has partly enabled us to achieve some objectives. Among those, the study of the kinetic compensation effect is one of the most desirable and demandable approaches for future generations.

3.2.7 Kinetics Compensation Effects (KCE) and its Role During Gasification

To understand the carbonaceous material/solid fuel gasification mechanism, a significant number of studies have been carried out on the aspect of change in char/biochar structure in different gasifying conditions [15–17]. Understanding the change in char/biochar structural properties with conversion correlating the char gasification path is very important. Also, those changes in char properties affected by gasifying conditions suggest the difference in the path of char gasification in different gasifying conditions. Based on the change in char/biochar structural behavior with conversion, the path of char consumption can be identified in different gasifying conditions. The outcomes of the studies were limited to specific experimental conditions, such as getting the actual regime with boundaries for a particular fuel, which will be very complex from the changes in char properties with conversion. However, previous studies have been conducted in TGA to obtain the reactivity data. Also, getting intrinsic char reactivity is essential for establishing the boundaries of the different regimes and specifying the path of char gasification. In addition, information on the intrinsic kinetics and instantaneous rate of product gas formation is essential from the aspect of the path of product gas formation with the understanding of their respective kinetics. Also, from an industrial perspective, continuous efforts have been made for a long time to maximize the yield of the desired product with higher char reactivity. In an overall approach, understanding char gasification and product gas formation mechanism are essential parts that provide the pathway to optimize the yield of desired products further during low-rank fuel gasification. Therefore, a systematic study approach is required to understand the low-rank fuel char gasification mechanism to define pathways of char gasification and product gas formation pathways. Studies on the generated kinetics (apparent values) during gasification are the prominent route to understanding the reaction mechanism. The char structure variation occurred due to char gasification, affecting reactivity significantly with conversion. Considering the past gasification studies of low-rank coal/other solid fuel and comparing those with the aspect of coal gasification technology, understanding the variation of apparent kinetic parameters (i.e., apparent activation energy and apparent frequency factor) with conversion is required. As noticed in previous works, the continuous variation in values of the apparent kinetic parameter for char consumption was expected, as a continuous change in char properties affects the gasification rate [10]. It is believed that generated apparent kinetic values would incorporate all the variable parameters, including the char structural properties, which will likely be affected by conversion. This can be expected for product gas formation, and overall, it will show the contribution of change in char properties with conversion to respective product gas formation. Given this status, there is a need to investigate and gain fundamental knowledge in low-rank fuel gasification. From the view of the product gas formation mechanism, some gasifying products are generally produced in most gasifying conditions. For example, the formation of CO_2 is a common gasification product during char-H_2O and char-O_2 heterogeneous reactions. However, the pathways of the reaction mechanism for the formation of CO_2 in these two gasifying environments may or may not be similar as the contribution of change in char structure of char-H_2O and char-O_2 reaction to their respective CO_2 formation might differ. This needs to be understood further from a fundamental aspect of the char gasification mechanism that has not been investigated in past studies of low-rank fuels.

As a result of the variation in kinetic parameters, the kinetic compensation effect phenomena were observed during the heterogeneous reaction of solid fuel with gases. Also, experimentally, it was proven that a proportional change in apparent activation energy (E_{app}) mechanistically resulted in a proportional change in apparent frequency factor (lnA_{app}) during gasification reactions. The proportional behavior represents a linear equation, as shown below (Eq. 3.2), where "m" and "c" are the slope and intercept of the kinetic compensation effect plot.

$$lnA_{app} = mE_{app} + c \tag{3.2}$$

In the study of the kinetic compensation effect, it was noticed that the change in rate during the gasification process resulted in a more significant/considerable change in E_{app} which is finally compensated by a change in lnA_{app}. The formation of proportional parameters and their interdependence are derived from rate equations. During the gasification of solid fuel char/biochar, the carbon present in solid fuel gets converted into CO, CO_2, and CH_4. Based on the conversion of carbon solid fuel/char/biochar into the carbon-containing gaseous product, the rate of char consumption (r_c), at a given conversion level (x), It is defined as the weight loss rate of coal char concerning moles of carbon species (i.e., moles of CO, CO_2, and CH_4) per unit mole of carbon remaining in the reactor. The rate of char consumption (r_c), which was calculated based on carbon, and the equation (Eq. 3.3) is presented below.

$$r_c = \frac{1}{W_c}\frac{dW_c}{dt} = \left(y_{CO} + y_{CO_2} + y_{CH_4}\right)\left(\frac{QP}{W_c RT_{amb}}\right) \tag{3.3}$$

Where y signifies the mole fraction, Q is the flow rate of fluidizing gas in L/sec, P = atmospheric pressure in bar, and W_c signifies the mole of char present in the reactor at any instant during gasification. W_c was calculated by subtracting the total amount of carbon released at any instant in time ($W_{char\ carbon\ released}$) from the total amount of carbon consumed during gasification ($W_{total\ char\ carbon}$). At any conversion level (x) with constant partial pressure, the rate of change in weight increase conversion was calculated by equating the rate of change in weight increase conversion to the Arrhenius equation, as shown in Eq. 3.4 [1]. Similar calculations for key product gas components can be done based on the gasifying environment.

$$r_c = \frac{1}{W_c}\frac{dW_c}{dt} = f(x)g(p)Aexp\left(\frac{-E_{app}}{RT}\right) = A_{char\ consumption}\exp\left(\frac{-E_{char\ consumption}}{RT}\right) \quad (3.4)$$

The study of simple kinetic and using kinetic compensation effects contributes to understanding the gasification mechanism. However, in the study of simple kinetic relative to study through kinetic compensation effects, it was noted that understating the reaction mechanism through investigating without compensation effect would draw limitations for identifying the value of the reaction mechanism. A study of the kinetic compensation effect would broadly identify the gasification mechanism and path of product gas formation. Knowing the trend of the kinetic compensation effect and the behaviors of E_{app} and lnA_{app} with conversion would add value to understanding gasifying reactor design, optimization of feedstock size, and process parameters. Finally, the implication would be identifying the path of key product gas formations. This will add value to the scaling up of gasification technology from bench scale to plant scale (Fig. 3.8a). On the other hand, studies carried out without kinetic compensation effects will be limited to finite points (Fig. 3.8b).

An in-depth understanding of the char gasification mechanism of low-rank fuel in different gasification environments will add value to industrial applications for optimizing the desired product gas concentration in the outlet stream of the reactor. Optimizing the concentration in the product stream could help in other processes,

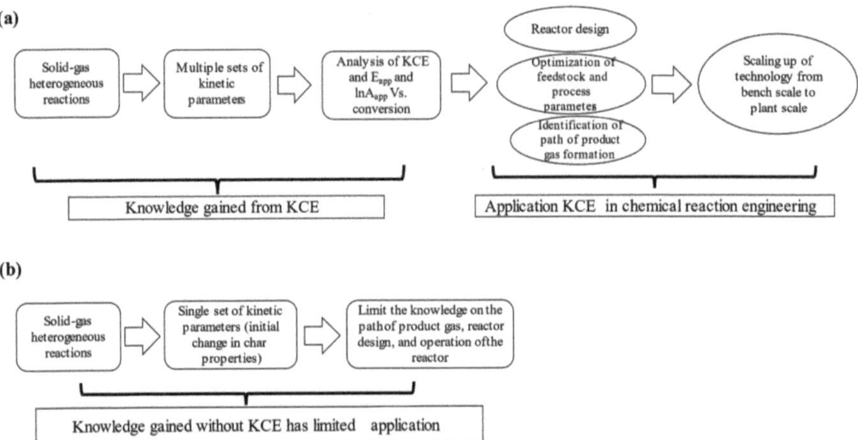

Fig. 3.8 Schematic flow diagram for the knowledge that can be extracted and its application in reaction engineering with KCE (**a**) and without KCE (**b**) [102]

for example integrating the industrial gasifier into the solid oxide fuel cell (SOFC). Hence, overall, the product syngas obtained from gasification can be effectively used to produce liquid chemicals and generate electricity through electrochemical conversion processes. The primary benefit will be the very low emission of CO_2 to the environment. However, some studies have continued to integrate the gasification plant into the SOFC station. The lack of complete understanding of the char gasification reaction mechanism with reaction kinetics limits the whole process for commercialization. Moreover, current efforts are being made to obtain the mechanism of co-gasification of coal and biomass and make it more beneficial for industrial use than single fuel. Therefore, having a complete understanding of each fuel participating in the co-gasification process will add value to a complete understanding of the mechanism of the co-gasification process. This will add deep insight into using low-rank fuel for the co-gasification process and help optimize the amount of any particular low-rank fuel needed to blend with other fuels. From the aspect of the catalytic effect inherent AAEM species, an in-depth understanding of the char gasification mechanism and the role of AAEM species on specific product gas formation would provide a platform to think further about the utilization of low-rank fuel char/biochar as the catalyst for the tar reforming process. Biochar has been used as a catalyst for reforming tar in CO_2 and H_2O environments [18]. However, the biomass has high volatile matter and less fixed carbon content than other low-rank coal. Also, among low-rank coal, some coal has a more disordered carbon structure that significantly differs from graphite structure, which could be used as catalyst support for reforming tar generated from other low-rank fuels.

3.2.8 Sources of External Parameters Affecting the Kinetics of Gasification Mechanism

The significant sources affecting the gasification mechanism are primarily divided into two categories. One is the inherent parameters of low-rank coal affecting the gasification rate, and the other is the Operational parameters and other evolved factors affecting the reactivity. A brief discussion on each category is presented below.

A. Inherent parameters of low-rank coal affecting the gasification rate

Oxygen Content The oxygen content in coal and char defines various functional groups, such as the phenolic and carboxyl groups, which help bind the metallic species inside coal. The reactive nature of the O-containing functional group contributes to and participates in the pyrolysis and gasification of coal [19–21]. Gasification of lignite coal from Zhundong coalfield (China) in CO_2 and H_2O environment displays a significant rise in the O-containing functional group with more condensation of char, i.e., condensation of minor aromatic ring present in char to bigger one [22]. Studies of Ye. et al [23] indicate that forming a new O-containing functional group and aromatic ring depends upon the pore opening with contact with the gasifying agent during char (Huating coal) gasification in the CO_2-O_2 atmosphere. Studies on Zhundong coal molten salt gasification suggest that the formation of O-containing functional groups during gasification are mostly carbonyl groups. Also, the diffusion resistance of CO and CO_2 inside the pore of char surpasses the carbonyl group [24]. The formation of the O-containing functional group and its concentration in char depend on the gasifying agent type. For example, the gasification of low-rank coal char in CO_2 significantly affects the concentration of the O-containing functional group than the gasification of char in H_2O. This is due to the difference in the structural aromatics ring condensation rate of char in different gasifying agents [25]. Studies of Zhundong coal char show that the peroxidation process of coal significantly impacts the post-process, i.e., char gasification [26]. This indicates that the formation of the O-containing functional group during pre-oxidation has a severe impact on enhancing char gasification and reducing the activation energy of char reactivity compared to the normal process. However, the formation of O-containing functional groups during pyrolysis mostly depends on their respective bond with char, as noticed from low-rank coal char pyrolysis. It was suggested that the O-containing functional group, such as 1,4-naphthalene dicarboxylic acid, poly (2,6-dimethyl-1,4-phenylene oxide, and poly 4-vinyl phenol) are released as primary pyrolysis products [27].

Alkali and Alkaline Earth Metallic (AAEM) Species Content Efforts are made to quantify different AAEM species in coal. The low-rank coal mainly contains sodium (Na), Calcium (Ca), magnesium (Mg), and potassium (K). On considering the catalytic activity of those minerals in heterogeneous gas-solid reactions, studies have also been done to investigate their catalytic effect during gasification reactions and in other reforming reactions (such as tar reforming) where coal char acts as a catalyst [28–30]. Past studies concluded that the catalytic role of inherent AAEM species during the coal gasification process depends on the chemical form of AAEM species in coal/char [31–34]. For example, the AAEM species (in the form of cation) present in coal could be associated with the carboxyl group present in coal char, and/or it can be in the form of a soluble salt such as NaCl [35–37]. This suggests that some low-rank coal could have higher mineral content, but it may not have a higher gasification rate than low-rank coal with relatively low mineral content. Past studies on coal char gasification suggest that the release of AAEM species such as Na and K during gasification varies with the gasification atmosphere. Also, it was observed that the release of Ca is higher during the gasification of char in CO_2 than H_2O and mixed H_2O/CO_2 atmosphere [38]. Studies also noted that the volatile AAEM species boned to the O-containing structure in char during volatile char interaction and enhanced char reactivity compared to AAEM-free char. This is due to the better catalytic effect of the inherent residual AAEM species in char [39]. This indicates that the catalytic role of inherent AAEM species depends on the chemical form of AAEM species in char. In addition, among various chemical forms, the catalytic activity of AAEM species is dedicated to certain chemical forms. The catalytic role of Na in the form of carboxylates in char (i.e., COONa)

NaCl has a significantly higher catalytic effect than NaCl in char [40]. Studies on Collie sub-bituminous coal suggest that O-containing structures contribute less to char-O_2 reactivity than variation in the ratio of smaller to larger aromatic ring systems. However, inherent AAEM species sub-bituminous coal has a less catalytic effect on char reactivity than lignite brown coal [41].

Moisture Content he moisture content of low-rank coal mined from different places is different due to various water content in the ground. The water content of Victoria brown coal from Western Australia usually goes to 65 wt.% as it is obtained in geographic regions that are very close to the sea [42]. Studies of low-rank coal indicate that the water present is completely molecular and can be removed by dewatering/drying [43, 44]. Also, water could be attached to the present O-containing functional group in coal, and the removal of this moisture could affect the physio-chemical properties of coal [45]. Hence, some studies have examined the effect of moisture content on coal gasification and the impact on the subsequent removal of water chemically attached to coal. The decrease in moisture content of low-rank coal significantly increases gas migration during char gasification. Also, it has a more significant impact on high-yield coalbed methane (CBM) development [46]. Hydrothermal dewatering from low-rank coal significantly impacts the properties and reactivity of coal during gasification [44]. It was noticed that at different temperatures, hydrothermal dewatering led to the removal of organic and inorganic elements at a different rate, as observed from past studies [47]. The inherent moisture content varies in low-rank coal; the inherent moisture content significantly impacts the adsorption capacity of methane in low-rank coal. This is because of the increase in hydrophilic and water-holding capacity of low-rank coal. The higher moisture content of sub-bituminous coal than other low-rank coal seems to be due to the modified graphite structure containing hydroxyl, carboxyl, and carbonyl groups, resulting in more water-holding capacity than other coals [48]. Studies on the abovementioned features pertinent to low-rank coal are very important, as char gasification plays a vital role in the overall gasification process.

B. Operational parameters and other evolved factors affecting the reactivity

To further understand the gasification process and mechanistic insights, rigorous studies have been carried out in the last decade. A gasification process usually consists of pyrolysis and subsequent gasification of char. In pyrolysis, a continuous release of volatile matter occurs from coal, leaving the char behind for gasification and other valuable processes. The released volatile matter, known as tar, contains a complex carbonaceous compound with some vapor of AAEM species. However, the release of AAEM species during pyrolysis is undesirable as it plays a catalytic role during the gasification process [29, 49]. Therefore, several studies have been carried out to observe the factors that influence the volatilization of AAEM species during pyrolysis of low-rank coal. Significant factors that could impact AAEM volatilization are temperatures, heating rate (slow heating rate and fast heating rate), pressure, and composition of the gas atmosphere surrounding the char particle [50]. To observe the actual effect of those operational parameters during pyrolysis, a wide range of experiments was designed using TGA, one-stage fluidized/fixed bed reactor, fixed bed reactor, and two-stage fluidized-bed/fixed-bed reactor [35, 51–53].

Depending upon the inert, reducing, and oxidizing gas atmosphere, AAEM species' volatilization varies, as observed in brown coal. The retention of AAEM species mostly depends on the reaction type and their participation in the reaction [54]. It was noticed that the role of AAEM species as an active site for adsorption and as a catalytic site leads to retention in char as AAEM spices are bonded to char. Therefore, in comparing different inert environments, i.e., He, and reactive environments, i.e., O_2, retention of more AAEM species occurs in O_2 with less volatilization than heated char in the He environment [31]. Hence, surrounding gas to heated particles affects the volatilization of AAEM species. Although the volatile-char interaction is very low in the case of the wire mesh reactor, the residence time and peak pyrolysis temperature could lead to significant volatilization of AAEM species, as observed from previous work on low-rank coal using the wire mesh reactor [31]. This suggests that the co-effect of particle time and temperature significantly affects the volatilization of AAEM species from coal during pyrolysis. The chemical form of AAEM species in coal and its valency also plays an essential role in its volatilization during pyrolysis. Also, it was evident from past studies that the monovalent AAEM species, such as Na, have higher volatilization capacity than divalent species like Ca and Mg, mostly in brown coals [55]. As observed from past studies of low-rank coal under certain experimental conditions, the pyrolysis of the NaCl-loaded coal and Na-form coal in a fluidized-fixed bed reactor [56]. It was noticed that the release of Na from NaCl-loaded coal, where Na is in the form of NaCl, is higher than that of Na-from coal, where Na is in the form of -COONa to char. In addition, the residence time of char under pyrolyzing conditions also plays a vital role in the volatilization of AAEM species, including the carrier gas velocity and bed height, which significantly impacts AAEM species' volatilization. The volatilized AAEM species in the gas phase are very reactive at a very high temperature. Hence, a high carrier gas velocity is preferred to avoid the re-condensation of back to the char. This is an important aspect to consider when designing an industrial reactor. However, the net release of AAEM species during volatilization is also influenced by the height of the bed because of the re-adsorption and desorption of volatilized species in the char bed. From particle size, the variation in particle size with an effect of intraparticle diffusion limitation could lead to significant variation in the release of AAEM species.

The above-discussed factor is based on the volatilization AAEM species during pyrolysis of low-rank coal, and consideration of these factors is vital in the gasification process. The char obtained from pyrolysis significantly contributes to the char gasification rate in the subsequent step. However, during char gasification, other factors (char conversion dependant factor), including factors generated from the experimental condition, affect the char reactivity. Past studies have been done to understand the impact of those factors on the enhancing/inhibiting char gasification reactivity during conversion. Based on the roles of AAEM species in catalyzing char reactivity, three major bulk parameters relevant to AAEM species could suggest the effect of inherent AAEM species on enhancing char reactivity. First is the chemical for inherent AAEM species in char during gasification. It was well known that Na has a higher catalytic effect among various AAEM species in low-rank coal than others. However, the actual impact of Na on catalyzing the gasification rate depends on its chemical form in char. The Na in coal char in NaCl form has a less catalyzing effect compared to Na in the form of Na_2CO_3 [57, 58]. It is because of the availability of Na as Na^+—that depends upon the affinity of Na^+ and other boned species to Na in char [59–61]. The second is the concentration of catalyst in char with conversion during gasification. Although there is a continuous increase in AAEM species concentration with the conversion, the impact of AAEM species concentration as a catalyst on the gasification rate is not linear with conversion [36]. Some other parameters affect the catalytic effect of AAEM species. Third, studies of low-rank coal suggest that the position of the catalyst inside the char matrix plays a vital role in participating in the gasification reaction. AAEM species on pore surfaces act as active catalytic sites as they can be easily accessible to gasifying agents. In contrast, the AAEM species are inside the char matrix. Hence, the catalytic AAEM species dispersion on char with conversion during char gasification has significantly influenced the gasification rate [62]. In addition, the continuous change in char properties (physical and chemical properties) with conversion affects the gasification char reactivity. Also, the properties of nascent char obtained from pyrolysis significantly affect the subsequent char gasification rate.

3.3 Kinetics and Reaction Mechanisms in Catalytic Gasification

The mechanism of execution of gasification reaction with and without catalytic effect are two different independent mechanisms. The path of product gas formation in a gasification reaction without any catalytic effect and with catalytic effect mainly differ in many gasification conditions. The variation in the path of product gas formation due to the catalytic effect also depends on the type of gasifying environment used. Also, the observance of the path of product gas formation is mostly speculated and determined in a kinetics-controlled regime because, in the kinetics-controlled regime, the behavior of the gasification reaction can be interpreted from intrinsic kinetics. This section elaborates on understanding gasification kinetics and reaction, emphasizing the catalytic effect of inherent AAEM species. The primary focus was given to the impact of key AAEM species, i.e., calcium (Ca), magnesium (Mg), sodium (Na), and potassium (K). In a broad aspect, this section explains the classification of catalysts and the role of the catalytic element in enhancing the gasification rate. Finally, the fate and roles of KCE during catalytic gasification will be presented.

3.3.1   Classification of Catalysts

The alkali and alkaline earth metal (AAEM) species significantly catalyze the gasification reactions/processes. Among the various AAEM species, Na plays a dominating role in catalyzing the gasification process. For an AAEM species to participate in/catalyze the gasification reaction, the retention of char at any particular conversion level and its chemical form inside the char matrix plays a vital role. The chemical form of AAEM species is characterized in two different aspects: one is on the base of origin, which is inherent to a solid fuel sample, and the other is based on minerals ion and their roles. An ample amount of investigation has been done to understand the catalytic activity of AAEM species during the gasification of char/biochar [12, 63, 64].

3.3.1.1   Based on Origin and Their Roles

Based on the origin, the AAEM species has a different chemical form. For example, in the case of Na-containing ions, the chemical structure of Na in the form of NaCl and Na2CO3 plays two distinct roles in catalyzing the gasification processes; it was noticed from an experiment that when Na-containing compound in the form of NaCl, less catalytically effective as compared to in the form of Na_2CO_3 [58]. This means the strong affinity between Na + and Cl- would inhibit the participation of Na + in catalytic processes. Hence, some effort has been made to convert the alkali halide into an active catalyst (using NH_3 and $Ca(OH)_2$) before using the solid fuel for gasification. However, this approach is suitable for academic lab scale analysis due to the higher cost associated with pretreatment processes. Contrary to this, some other methods have been made to make the AAEM species a catalyst for the gasification of biochar/char. This can be achieved by optimizing the pyrolysis process, which is the pyrolysis heating rate (slow or fast pyrolysis) and volatile char interaction [62]. It is possible to achieve a favorable outcome for the preferential release of Cl [35].

The experimental study performed on a lab scale on raw brown coal and NaCl-loaded brown coal shows the different trends in behaviors for releasing Na and Cl during fast and slow pyrolysis (Fig. 3.9). In all the pyrolysis cases, Na and Cl release is not in the form of NaCl. Mainly, the release/volatilization of Na and Cl occurs individually. The raw coal sample shows a favorable result in the fast release of Cl with an increase in temperature, which is desired to occur during slow pyrolysis operation. Meanwhile, in the fast pyrolysis process, a decline in the release of Cl occurs with an increase in pyrolysis temperature, and continuous release of Na occurs with an increase in temperature.

Fast pyrolysis shows a non-monotonous trend in the release/volatilization of Cl and Na with the rise in temperature in the case of a loaded coal sample. This suggests that the chemical forms of inherent species are different based on the fuel sample and its origination. If a coal sample is collected from sea level, such as Loy

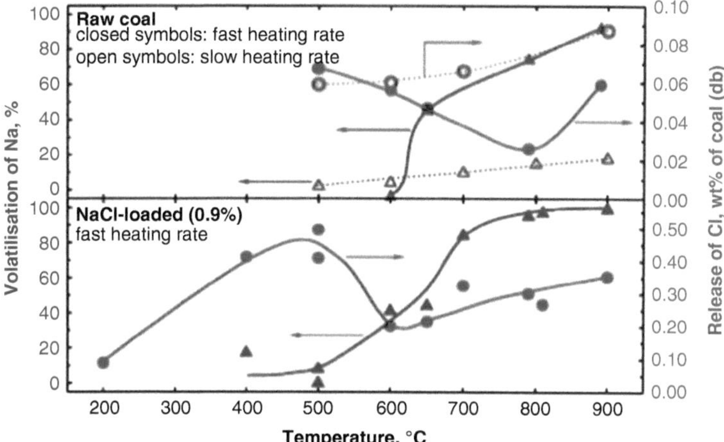

Fig. 3.9 Volatilization of Na and Cl during the pyrolysis of Loy Yang raw coal and NaCl-loaded (loading level: 0.9 wt%) coal samples in the one-stage fluidized-bed/fixed-bed reactor when the coal particles were heated up rapidly ($>10^3$ K s^{-1}) or slowly ($\sim$10 K min^{-1}) [30]

Yang Victorian brown coal, Australia, the moisture content is salt water—that responds to the presence of Na in NaCl. Therefore, before defining a fuel sample for a gasification process by emphasizing the catalytic effect of inherent AAEM species, attention needs to be given to its origin and the inherent chemical form of AAEM species.

3.3.1.2   Based on Minerals Ion and Their Roles

In the process of gasification of solid fuel in different gasifying environments. The retention of AAEM species increases with the increase in char conversion. The retention of AAEM species (Na, Mg, Ca, and K) in char with a change in char conversion is shown in Fig. 3.10. It can be noticed that a significant increase in the concentration of AAEM species in char with increased char conversion, particularly to a gasifying environment containing steam (15%H_2O-Ar and 15%H_2O-CO_2). However, there is some increase in the retention of AAEM during the gasification of char in a pure CO_2 environment, but the concentration is not as high in the steam environment. This indicates that chemical form and mineral ions achieved during char gasification play a vital role in catalyzing gasification. This suggests that during the gasification of char in a CO_2 environment, the volatilization is remarkably high due to the formation of carbonate in the char matrix, leading to less holding capacity of AAEM species [65, 66]. It can be suggested that the catalytic activity of AAEM species varies with the gasifying environment and depends on the proper chemical form and the AAEM metal ion to be held by the char matrix.

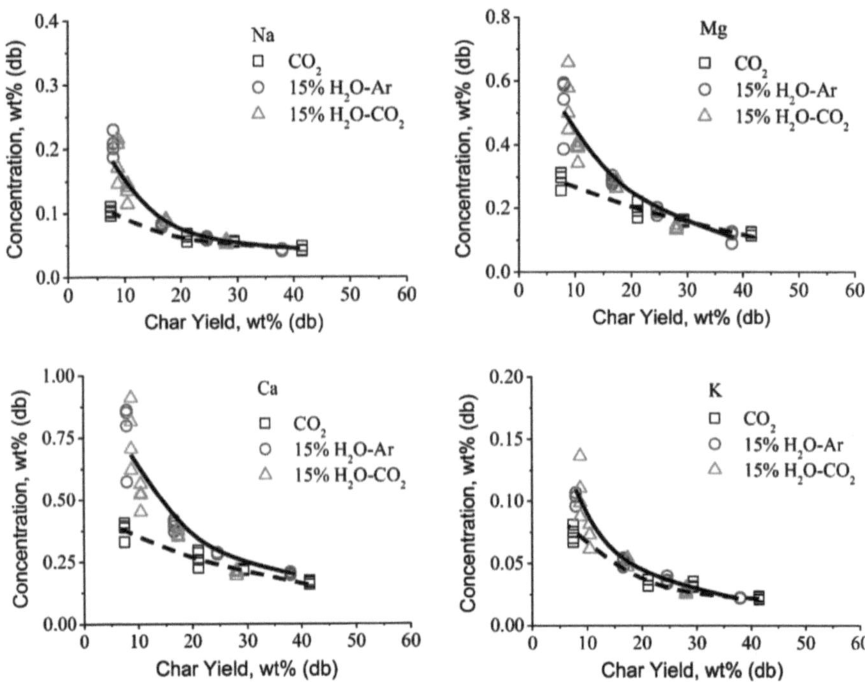

Fig. 3.10 Concentrations of AAEM species in char versus char yield for the chars obtained in pure CO_2, 15% H_2O balanced with Ar and 15% H_2O balanced with CO_2 at 900 °C during the gasification of Collie sub-bituminous coal [41]

3.3.2 Role of the Catalytic Element in Enhancing the Gasification Rate

There are certain ways via which the catalytic effect of AAEM species is to be forecasted to enhance the gasification reaction, and any of those steps could be the rate-limiting step for the gasification process. A gasification process can be improved by accelerating the rate-limiting step. The possible ways in which AAEM can play a dominating role in catalyzing are as follows:

- Adsorption of gasifying agents and chemical desorption of those on the char surface.
- Based on the fates of radicals on the char surface, the discouragement for radicals (through the chemisorption of a gasifying agent) leads to further participation in the gasification reaction.
- By participating in the reaction between radicals derived from an adsorbed gasifying agent and char matrix.
- By proving a low-energy environment for the desorption of gasifying agent from char surface [67].

As mentioned above, any of those steps could potentially be a limiting step and equally responsible for speeding up the rate of gasification reaction. However, identifying the appropriate rate-limiting step is one limiting step for scientists and engineers in the engineering and industrial community. Considering the primary steps for executing a chemical reaction, such as adsorption, surface reaction, and desorption, some explanations will be presented in the section to understand the gasification process better.

3.3.2.1 Adsorption of Gasifying Agent on the Char Surface

The adsorption of the gasifying agent on the char surface usually occurs on both carbon active sites and catalytic active sites. The adsorption mechanism usually differs for gasifying environments and varies with the regime. Specifically, based on an adsorption process such as adsorption on a catalytic site, the adsorption step is followed by surface diffusion and then further followed by surface reaction and desorption. However, in the case of direct adsorption on the carbon active site, it led to surface reaction and desorption without any step occurring for surface diffusion. Moreover, the path of the respective gas formation may vary with different catalytic and carbon active site adsorption. It can be said that based on the adsorption active site, product gas composition for a gasification reaction may vary based on changes in the path of product gas formation. As experimentally studied for char gasification in a limiting oxygen environment, it can be seen that the dissociative adsorption of oxygen on the char surface on both catalytic and carbon-active sites leads to two different gasification mechanisms. It can be seen that the adsorption of oxygen on the catalytic active side leads to surface reaction and radical formation and subsequently results in the formation of CO (Fig. 3.11a). Whereas the dissociative adsorption of oxygen on the carbon active site leads to a surface reaction with nearby adjacent occupied carbon active sites, following the formation of CO_2 occurs (Fig. 3.11b). This suggests that product gas formation is partly attributed to the type or kind of dissociative adsorption that arises. Although this kind of speculation could vary with different types of gasifying environments, it can be stated that changing the adsorption mechanism substantially impacts the path of the product gas formation mechanism.

3.3.2.2 Surface Reaction and Diffusion of Intermediate to Inner Char Matrix

The surface reaction and diffusion of intermediate species are, in some cases, interdependent and, in some cases, are not. As can be depicted from Fig. 3.12a–b, the surface reaction primarily occurred between the non-dissociative adsorption of H_2O (steam molecule) and CO. CO is formed through the dissociative adsorption of steam on the carbon surface. Non-dissociative steam adsorption occurred on a catalytic active site on the surface. The surface reaction between H_2O and CO primarily

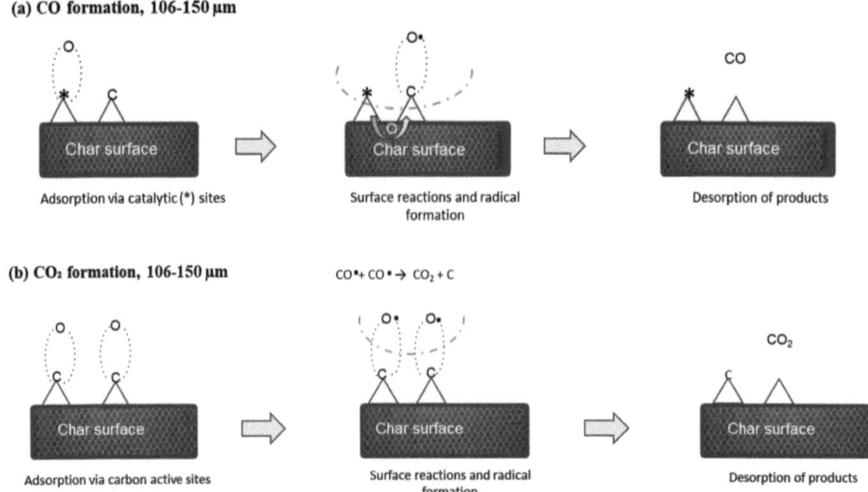

Fig. 3.11 Reaction mechanisms CO (**a**) and CO_2 (**b**) formation during gasification of low-rank fuel char in 0.4%O_2-Ar gasification environment [2]

leads to surface water-gas shift reactions, specifically observed in the case of particle size 106–150 μm. However, in the case of bigger particle size (180–212 μm), the formation of CO inside the char matrix got re-adsorbed on the char surface. Those re-adsorbed CO on the catalytic active site further participate in surface reaction by combining it with the adjacent adsorbed H_2O molecule—that finally leads to the formation of CO_2 and H_2 through a catalyzed water gas shift reaction. Therefore, it can be suggested that based on the formation of surface radical, the respective surface reaction and path of product gas formation are different, leading to other prospects for surface and catalytic water gas shift reactions.

3.3.2.3 Desorption

The desorption mechanism is the subsequent step of the surface reaction. This step is the execution of product gas formation. In the case of catalytic and non-catalytic desorption, the available active site could be a free catalytic active side or another non-catalytic active site. In some cases, the desorption step is usually accelerated by the inherent catalyst present inside the char. In very rare cases, it was found that the desorption step is the rate-limiting step of the overall processes, as most of the desorption occurs much faster than the other two adsorption and surface reactions. The active site left behind after desorption and subsequent adsorption of gasifying agents on those sites can be monotonous processes for controlling the overall gasification process. However, during a gasification process, the continuous consumption of carbon with the formation and consumption of active sites results in the process's non-linear and non-monotonous nature. Therefore, it would be so interesting to say

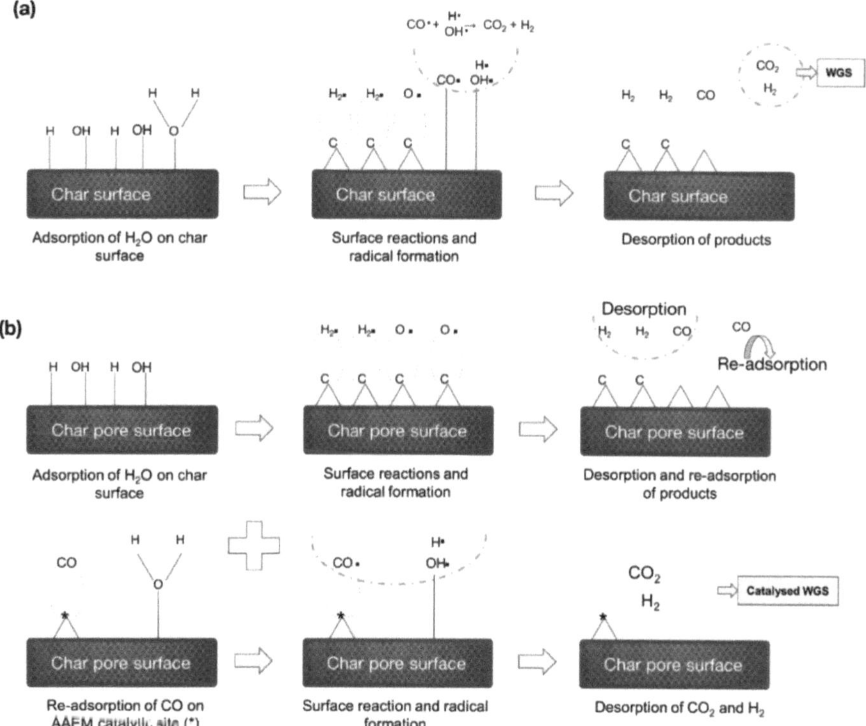

Fig. 3.12 Mechanistic pathways for steam gasification of low-rank coal for particle sizes 106–150 µm (**a**) and 180–212 µm (**b**) [1]

that though a certain number of chemical reactions occur during gasification, the execution of those chemical reactions is different and away from the occurrence of individual chemical reactions.

3.3.3   Fates and Roles of KCE During Catalytic Gasification

During gasification of low-rank fuel, with a change in particle size, the retention of AAEM species in char varies a lot. The gasification of char in the higher range of particle size is porous due to the high release of volatile matter. At the same time, the bigger size char high also had high AAEM species content. Mechanically, it was noticed that the porous nature and high AAEM content in more significantly bigger-sized char than smaller-sized char led to differences in char gasification and product formation mechanisms. This means there are enormous possibilities of differing product gas formations during the gasification of bigger-sized and smaller-sized chars. In brief, it can be suggested that a catalytic mechanism occurs during char

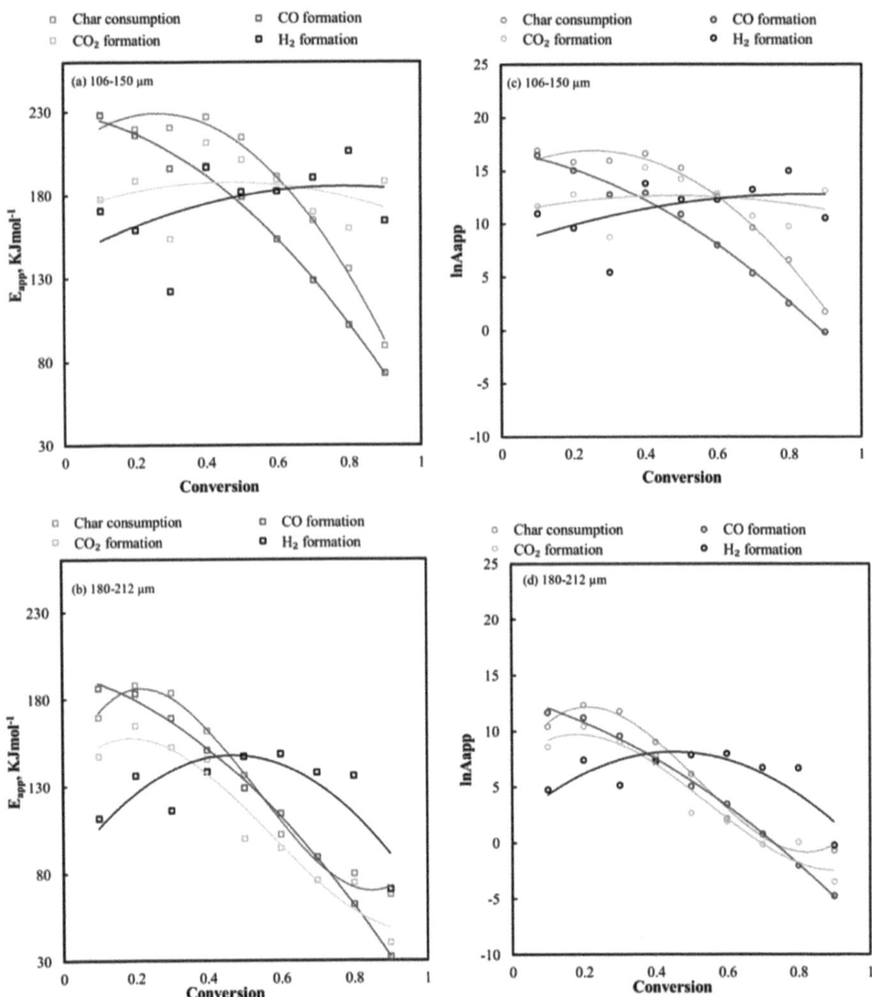

Fig. 3.13 Variation of gasification kinetics (lnA_{app} and E_{app}) with conversion for particle sizes 106–150 μm (**a & c**) and 180–212 μm (**b & d**) in a steam environment (15%H_2O-Ar) of low-rank coal [1]

gasification, substantially contributing to any step of the gasification reaction (adsorption, surface reaction, and desorption), leading to a significant alteration in the key product gas formation pathway.

The lab scale studies done for the gasification studies of two different particle sizes (106–150 μm and 180–212 μm) in steam environments differ in lnA_{app} and E_{app} at any particular conversion level, as it can be seen from Fig. 3.13a–b, for smaller sized char a continuous decrease in the kinetic parameter of char consumption, CO_2 formation, and H_2 formation. The kinetic parameters of CO formation become

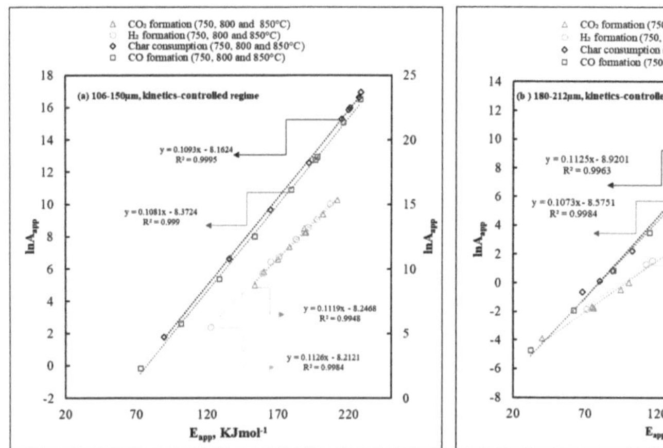

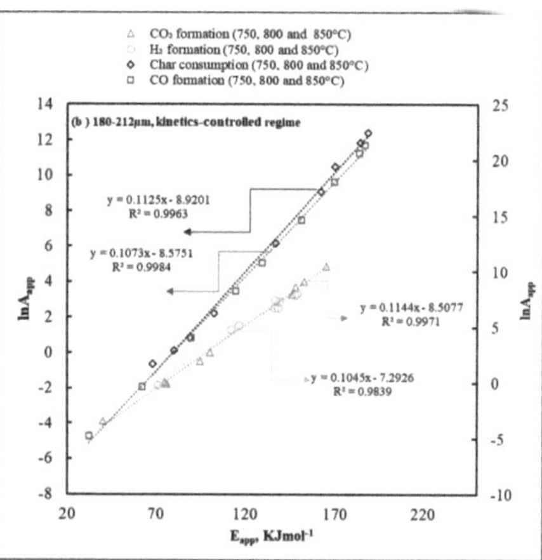

Fig. 3.14 Kinetic compensation effect plot for char consumption and product gas formation (CO, CO_2, and H_2) for particles of 106–150 µm (**a**) and 180–212 µm (**b**) during gasification of char in steam environment kinetics controlled regime [1]

constant at the higher conversion level. On further investigation of the kinetic compensation effect plot Fig. 3.14a, it can be mechanistically stated that H_2 and CO_2 take place from heterogenous char-H_2O reaction with the decrease in apparent activation energy and apparent frequency factor with the increase in conversion. Furthermore, intraparticle diffusion limitation was noticed in the case of bigger particles, which showed a low value of gasification kinetics (lnA_{app} and E_{app}) compared to smaller-sized char at all conversion levels. Although the operations for both particles are carried out in a controlled regime, intraparticle diffusion limitation was attributed to forming a more porous nature due to the release of volatile matter. In the case of bigger particles, the change in kinetic parameters of char consumption and formation of CO, CO_2, and H_2 by passing through the peak value near to mid of the char conversion (Fig. 3.13c–d). After closely analyzing the kinetic compensation effect plot, it was noticed that (Fig. 3.14b) CO, CO_2, and H_2 are produced from the heterogeneous reactions between the char surface and H_2O.

Finally, it can be speculated that understating the variation of gasification kinetic with conversion would partly allow us to understand the gasification mechanism. However, the cross interrelation between variation of kinetics and kinetic compensation effect would enable us to draw a solid conclusion on understanding the gasification mechanisms. The slope, intercept, and extent can be interpreted from the kinetic compensation effect plot. Understanding the accurate kinetic compensation effect mechanism would allow us to use the suitable plot form where we can compare different solid fuels based on their changes in physical and chemical properties, which would also be inherent to them during the process of char gasification.

## 3.4	Gasification Thermodynamics

Solid fuels used for gasification have a wide range of variations in physical and chemical properties. This variation significantly impacts gasification conditions and product quality. Considering those aspects, limitations, and conditions of the gasification mechanism, understanding gasification thermodynamics is usually a theoretical approach to critically observe how the different properties of solid fuel, operation temperature, pressure, and gasifying agent to solid fuel ratio impact each other during gasification [68]. When examining the theoretical limits of the chemical species distribution at any particular equilibrium level, there are certain limits to the thermodynamic analysis of gasification reaction. Moreover, approaches have been made to assessing the thermodynamic efficiency for any given solid fuel and gasifying condition. In general, thermodynamic predictions have inherent limitations, and past efforts have been made to play with gasification kinetics under different conditions.

### *3.4.1	Thermodynamic Analysis for the Gasification Process*

Specifically, a chemical equilibrium for a gasification reaction is usually obtained by maximizing the entropy and minimizing the Gibbs free energy of the system. In general, two kinds of approaches have been made for developing a thermodynamic equilibrium: one is stoichiometric, and the other is nonstoichiometric. The stochiometric approach is needed to define the chemical reaction of all participating chemical species. In contrast to that for a non-stoichiometric response, the elemental composition of solid fuel (C, O, H, S, and N), as received from the ultimate analysis of that particular sample, was fed to the theoretical investigation system. The non-stoichiometric analysis is appropriate for understanding gasification reactions with uncertain mechanisms for solid fuel whose chemical structure and composition are unknown. The non-stoichiometric approach is beneficial, but there are some deviations between experimental and theoretical investigation through the thermodynamic approach. The deviation is due to adequate assumptions such as equilibrium conditions, consideration of char and tar as part of solid carbon, and treating the ash as an inert species. Much work has been developed to address those problems, and strategy has been made recently.

#### 3.4.1.1	Gasification Reaction Equilibria

The prediction and development of thermodynamic modeling of biomass gasification is very complex. However, it could be partly linear for each feedstock. However, developing all solid fuels that could be used for gasification in generic terms is very complex. This is because each individual solid fuel's physical and structural characteristics differ. For example, a marginal variation in physical properties was noticed when comparing coal with biomass. In another aspect, the vast experience obtained from coal and other carbonaceous materials can be implemented for the advancement of the future in the development of sustainable gasification technology by adopting a wide range of feedstock such as biosolid, solid waste, and other renewable solid sources. Basically, solid fuel gasification involves complex reactions by interacting between the solid and gas phases. Those reactions include pyrolysis, partial oxidation, steam gasification, and other secondary gasification reactions. A thermodynamic equilibrium model is considered a zero-dimensional model as the modeling approach is irrelevant to the design of the gasifier. The prime objective of thermodynamic equilibrium modeling is to achieve a high solid conversion yield with optimal processing conditions. The determination of optimal process conditions also provides information on the residence time of the feedstock inside the reactor at an equilibrium state [69]. Based on the feed elemental composition, the overall procedure to approach simulating a gasification process using the equilibrium kinetic model approach, i.e., non-stochiometric approach [70, 71] (Fig. 3.15) and stochiometric approach (Fig. 3.16) [72–74], is discussed briefly in the following picture.

3.4.1.2 Phase Equilibrium

In general, gasification processes with longer residence time operated at higher temperatures (> 800 °C) are suitable for the equilibrium model. The equilibrium model assumes that reactants are thoroughly mixed and the reaction time is very long enough that the gasification bed is very stable with no further reaction to occur. Based on the previously developed approach, the thermodynamic phase equilibrium approach is the methodology that is mainly implemented. There are some

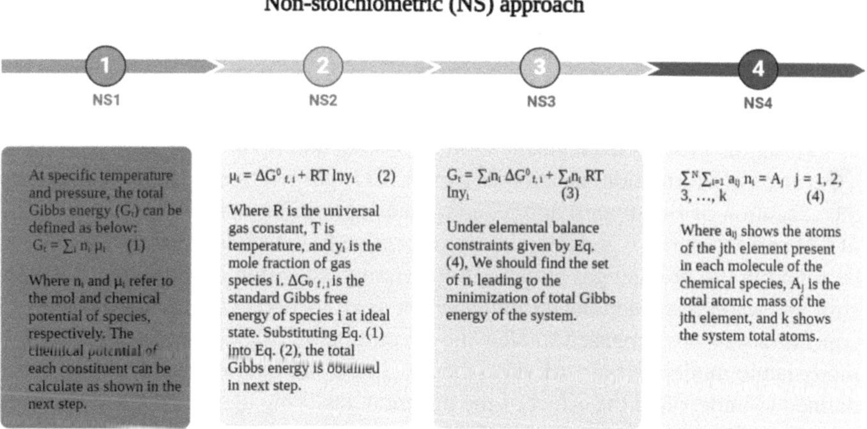

Fig. 3.15 Non-stochiometric (NS) approach for gasification reaction equilibria [70, 71]

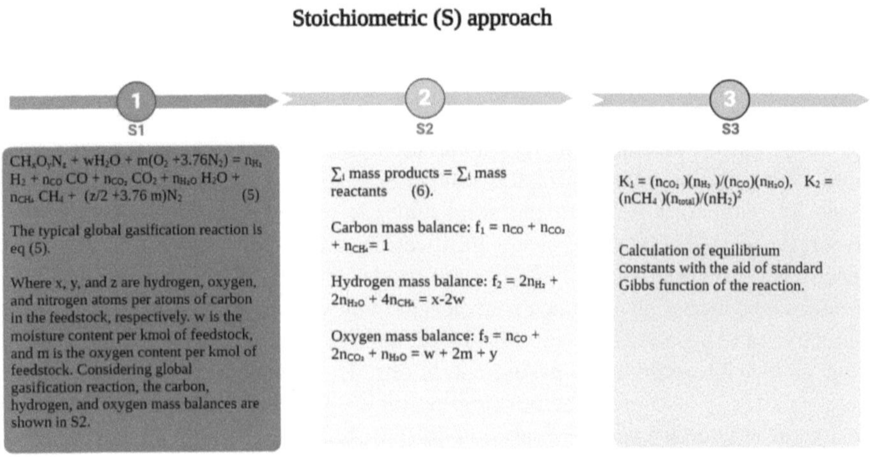

Stoichiometric (S) approach

S1

$$CH_xO_yN_z + wH_2O + m(O_2 + 3.76N_2) = n_{H_2} H_2 + n_{CO} CO + n_{CO_2} CO_2 + n_{H_2O} H_2O + n_{CH_4} CH_4 + (z/2 + 3.76 m)N_2 \quad (5)$$

The typical global gasification reaction is eq (5).

Where x, y, and z are hydrogen, oxygen, and nitrogen atoms per atoms of carbon in the feedstock, respectively. w is the moisture content per kmol of feedstock, and m is the oxygen content per kmol of feedstock. Considering global gasification reaction, the carbon, hydrogen, and oxygen mass balances are shown in S2.

S2

$\sum_i$ mass products = $\sum_i$ mass reactants   (6).

Carbon mass balance: $f_1 = n_{CO} + n_{CO_2} + n_{CH_4} = 1$

Hydrogen mass balance: $f_2 = 2n_{H_2} + 2n_{H_2O} + 4n_{CH_4} = x - 2w$

Oxygen mass balance: $f_3 = n_{CO} + 2n_{CO_2} + n_{H_2O} = w + 2m + y$

S3

$K_1 = (n_{CO_2})(n_{H_2})/(n_{CO})(n_{H_2O})$, $K_2 = (nCH_4)(n_{total})/(nH_2)^2$

Calculation of equilibrium constants with the aid of standard Gibbs function of the reaction.

Fig. 3.16 Stoichiometric approach (S) for gasification reaction equilibria [72–74]

significant advantages of the equilibrium model over another approach, which make it straightforward to implement and allow us to get a speedy converged solution compared to the kinetic model. However, it should be noted that the equilibrium model is used to design the reactor. The equilibrium model is also achievable to get product gas composition only at the outlet of the reactor. In contrast, a kinetic model can predict the gas composition at any point inside the reactor. Since every model cannot be perfect as this is based on certain assumptions, here are some pros and cons of thermodynamic equilibrium models presented in the table below (Table 3.3).

3.4.2   Thermodynamic Properties of the Gasifying Fluid

From the aspect of costing a gasification operation, the use of air in gasification industries is more desirable. However, using air promotes the production of low heating value syn gas. Adding pure oxygen to a gasification process could significantly alter the equilibrium concentration. However, the rise in O_2 also adds value in shifting the process from gasification to combustion [69]—that causes more CO, CO_2, and H_2O formation. In another aspect, CO_2 gasification is quite helpful from the execution of Boudouard and reverse water-gas-shift reaction; the concentration of CO and H_2 increase and decrease in respective to their rate. Also, more CO_2 can be found in the product gas stream. In the stem gasification reaction, the water gas shift and water gas reactions introduce water as an oxidizing agent. The H_2 and CO_2 concentrations are expensed to alter the accumulation of CO content. From a thermodynamic modeling point of view, the specification of various oxidizing agents is defined to understand the effect of the different gasifying agents in thermodynamic modeling. The specification of different gasifying agents is called the gasifying fluid's thermodynamic properties, which is shown in Fig. 3.17.

Table 3.3 Advantages and disadvantages of thermodynamic models [69, 72, 75–79]

Advantages	Disadvantages
Gasifier geometry is usually not considered, allowing the model's applicability in a wide range	Not suitable for low gasification operation temperature as a limitation for achieving the thermodynamic equilibrium
Knowledge of the conversion mechanism is not required as it focuses on reaction residence time and operation temperature	Obtained results are not especially precise for direct adoption for an industrial-scale operation
All to set operation conditions without any limitation	It is challenging to account for and identify the char and tar formation mechanism
The utmost conversion of the reactants and maximum product yield at particular conditions can be obtained	Possibilities in getting over estimation for H_2, CO, and product yield. Underestimation of CO_2 and CH_4 generation
An analysis of the sensitivity of another operating parameter can be performed	Limited to the hydrodynamic analysis of the thermal operation as the geometry and design of the gasifier are not considered
Very user-friendly and easy to implement with any gasification processes	Difficult to achieve accuracy in the estimation of tar content
Less computational time and not very computation intensive compared to another model and optimal operating conditions for a gasification process can be obtained	CH_4, tars, and char composition are generally not considered at the outlet stream

3.4.3 Application of Thermodynamics in Gasification

The application of thermodynamics is quite diverse in terms of the development and use of thermodynamic modeling. Although specific applications exist, some are based on studying parametric conditions to optimize the process condition. The major approaches that have been made to understand some dynamic behaviors and the influence of change parametric conditions on the overall process are listed below.

- Effect of temperature in the gasification process thermodynamically.
- Effect of solid fuel and gasifying agent equivalence ratio during continuous operation of gasification processes.
- Effect of gasification agent on gasification kinetic and yield of the desired product inside any point of the reactor and at the outlet of the reactor.
- Effect of moisture content of inherent moisture content on overall gasification process and yield of desired product gases.

In addition to the above, thermodynamic analysis is used to determine the energy values of a gasification process. This includes evaluating and assessing the syngas energy values, tar energy values, and exergy values of the thermochemical conversion processes [80].

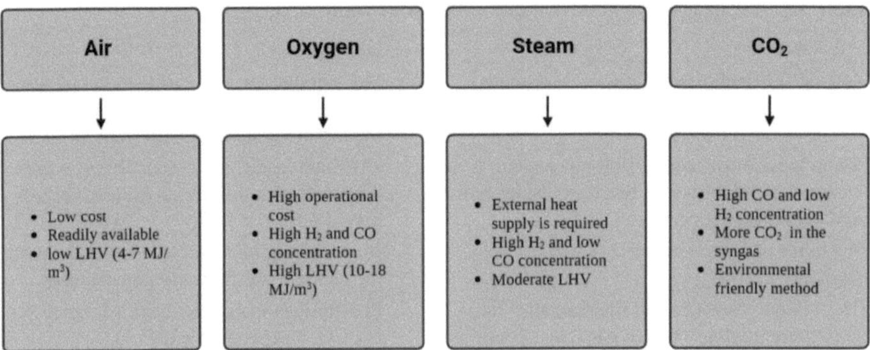

Fig. 3.17 Specifications of various oxidizing agents for thermodynamic modeling approach [69]

3.5 Transport Phenomena in Gasifying Beds

The transport phenomenon is an underlying process that occurs in every thermo-chemical process. The transport phenomena include heat, mass, and momentum transport occurring during the execution of the thermochemical conversion of solid fuel, and this provides momentum, heat, and mass transport. Studying transport phenomena and observing the underlying behavior experimentally in some cases is quite challenging as gasification operation is carried out at very high temperatures, and safety/hazard usually does not allow this objective. Therefore, cold flow beds and computational approaches have been made to investigate and observe those behaviors. They can enable us to forecast the process variation from lab to pilot scales. A brief explanation of each transport during the gasification operation will be given in the following sections.

3.5.1 Momentum Transport in Dynamic Gasifying Beds

The momentum transfer, such as gas-solid and solid-solid interaction, is usually noted during the dynamicity of the gasifying bed. The dynamicity of the gasifying bed is generally indicated in observed cases. Fluid dynamics plays one of the primary roles in affecting the gasification processes. The dynamic gasifying bed includes a fluidized bed, an entrained bed, and other moving beds, etc. [81]. In this chapter, a schematic of fluidized bed gasification is presented for understanding the momentum transfer mechanism [82, 83]. However, identifying the momentum transfer mechanism for the gasification process alone is very rare. This is because the impact or momentum transfer is not always limited to a particle or close bound-ary region. The momentum effect is also shown by other sub/interdependent pro-cesses, such as heat and mass transfer. Schematically, a circulating fluidized bed gasifier is shown in Fig. 3.18. The bed contains two different solid particles (bio-mass and solid particles of marginal variation in density. The inherent parameters

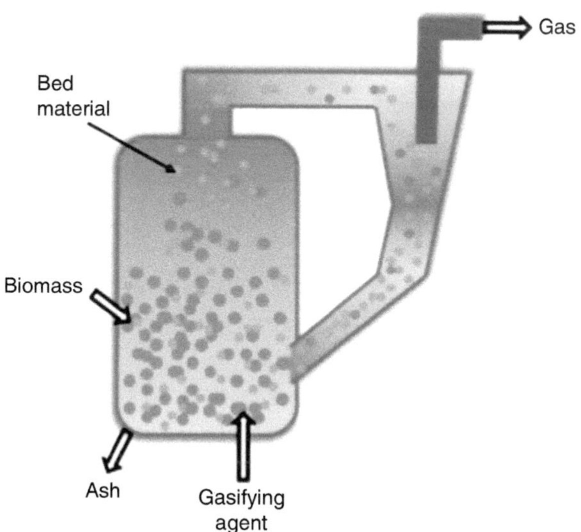

Fig. 3.18 Schematic of a circulating fluidized bed (CFB) gasifier [82]

and process parameters, such as the density of particles, the flow rate of gas to gasification reaction, and the higher diameter of the gasification reactor, play a crucial role in creating and affecting the momentum interaction in a gasifying bed. Understanding the hydrodynamic behaviors of the gasifying bed from the aspect of momentum transfer would partly enable us to see whether an immense mixing of gasifying agent and solid fuel is occurring.

3.5.2   Heat Transport in Gasifying Beds in Non-Isothermal Systems

Many more modeling approaches have been made to observe the heat transfer phenomena. The current section presents a multiscale modeling approach to see the fluid dynamic behaviors impacting the heat transfer mechanisms [81, 84]. In principle, during a gasification operation, the initial impact of heat transfer occurs at particle scale on subjecting inter and intraparticle heat transfer effects/limitations. Then, the variation occurs at the ongoing affect reactor scale, which can be visualized through variation in fluid dynamics. This means the boundary condition (temperature and gas composition) is assigned to the particle level and expanded to the reactor level. Finally, the combination of observations made on particle and reactor levels will add value to visualizing the impact at a multiscale level. Findings made from a multiscale level have a broader impact in retrospect, a level that is not limited to lab-scale reactors and a commercial scale (Fig. 3.19) [85]. Moreover, understanding the gasification phenomena from the aspect of the heat transfer mechanism is always going to solve all engineering and science-related problems, as heat, mass, and momentum transfer are interrelated and equally important to each other.

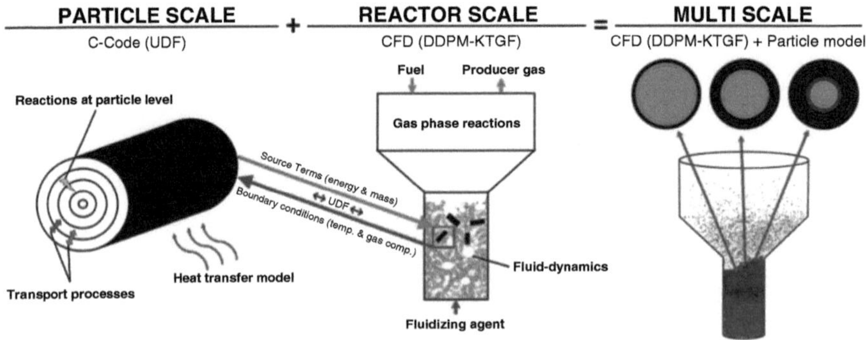

Fig. 3.19 Multiscale modeling of fluidized bed gasification [85]

3.5.3 Mass Transport in the Gasifying Bed

At a particular level, solid fuel gasification has two kinds of diffusional mechanisms (mass transfer is usually noticed). The basic approach is given to gas-solid interaction to understand the mass transfer mechanism occurring in any gasification process, irrespective of the gasifying reactor. Gas-solid interactions are basically considered to emphasize the gasifying agent and solid fuel interaction [81, 86]. However, as stated, gas-solid interaction during gasification is not as simple as the two kinds of mass transfer mechanisms that were broadly noticed experimentally and theoretically in the previous studies. Those are the intraparticle diffusion mechanism and interparticle diffusion mechanism. As per the depicted image of the char-CO_2 gasification mechanism (Fig. 3.20), the intraparticle diffusion mechanism is closely related to the function of particle size. The intraparticle diffusion mechanism explains the diffusion of gaseous components inside char/solid particles through pores (also known as the internal diffusion mechanism). This mechanism is related to the interaction of a single particle in the gaseous phase in a gasification operation. Therefore, any particle's influential ness factor ($\eta_{intraparticle}$) is a function of mean diameter (mean) [87]. In the case situation in an actual gasification environment, in addition to the effect of intraparticle diffusion, the interparticle diffusion limitation was also analyzed as its main phenomenon and the mechanism underlying the gasification bed. This interparticle diffusion mechanism explains the diffusion of gasifying agents inside solid porous beds. Therefore, over interparticle diffusion/effective diffusion of gas components inside the bed is a function of the particle size distribution (PSD) in that particular bed.

Overall, it was speculated that the effectiveness factor at the particle level (intraparticle diffusion factor) decreases with an increase in particle size. However, the effectiveness of a factor of bed (interparticle diffusion factor) increases with an increase in the particles [88]. This is because, with an increase in the particle size of the bed, the path between particles becomes less porous, less surface area is available for interaction, and more gaseous material will be available for particle-to-particle diffusion.

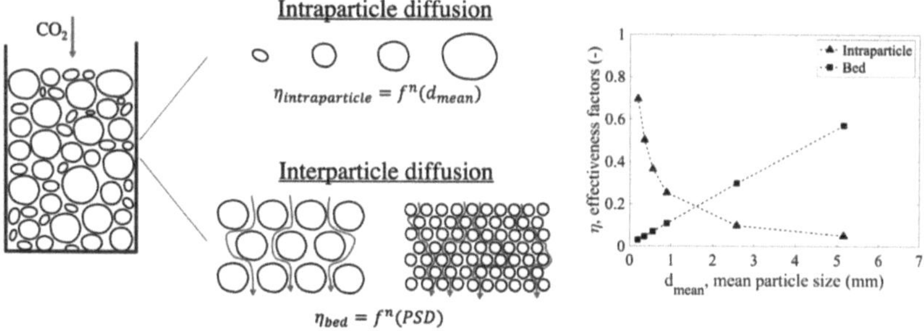

Fig. 3.20 Schematic view of intraparticle and interparticle diffusion during CO_2 gasification of biomass char in a packed bed [87]

3.5.4 Role of Thermal Conductivity for Energy Transport in Gasifying Systems

Gasification is the thermochemical conversion of solid fuel in a fluid/gasifying environment. The heat transfer from solid to gas, gas to solid, and solid to solid is a primary role in executing the heat transfer in bed. Heat transfer in a gasifying bed usually occurs on a molecular scale. The heat transfer from gas to solid is essential in a continuous gasification operation because the gasifying agent acts as a carrier gas for carrying heat to the solid bed [89]. Therefore, a higher thermal conductive gas is always desirable. However, looking at the constraints of inter-pyrolysis gas and particular gas for acting gasifying agents, the reactor's gasification operating condition and design play crucial roles in efficient operation. On emphasizing the operating temperature, it can be stated that the gasification temperature is most important as the thermal conductivity of gas is the function of temperature. In alignment with the design of a gasification reactor, it should provide a high contact between gas and solid. Finally, it concludes that the reactor's design plays a vital role in subjecting to the conduciveness of the gasifying agent [90]. It should be noted that while optimizing the process or developing a model of a gasifying process and reactor, the functional variation of the thermal conductivity of gasifying agent should be included.

3.6 Advances and Research Progress in Engineering Aspects of Gasification Technology

Advancements in research on gasification technology are continuous and monotonous. Enhancing the gasification process and developing more robust technology have been widely expanded in every aspect of engineering. Considering the recent demand for gasification technology, primary attention has been given to the advancement of gasification kinetics, thermodynamics of gasification technology, and the role of transport phenomena during gasification processes. The current section of the chapter presents a brief discussion on the advancement of each side of the engineering of gasification technology.

3.6.1   Current State of the Art in Gasification Kinetics

Gasification kinetics has played an essential role in optimizing the process parameters for highly designed product yield, designing a gasification reactor, and attribution of different solid feedstock for thermal operation. In subjecting the fundamentals of reaction kinetics, many kinetic model approaches have been made, such as single-step and multi-step kinetic models. The selection of an appropriate model depends on the trend of experimental information available [91]. There are some challenges associated with developing a kinetic model with some assumptions. In some cases, the challenges are defined clearly. Understanding the implicit consideration is essential, and reviewing the implicit assumption would improve the kinetic analysis [92]. Those assumptions are related to mass transfer limitation, char synthesis, and the existence of a maximum reaction rate.

- Mass transfer limitation: It was assumed that below 1000 °C, the gasification reaction is controlled by chemical kinetics (kinetics-controlled regime), and mass transfer has no significant impact on the processes. However, a recent study has speculated that intraparticle pore diffusion has been playing an important role due to changes in particle size. This cannot be overlooked. However, the intraparticle diffusion mechanism is independent of the geometry of the reactor configuration.
- Char/biochar synthesis: while developing a kinetic model of a process, the char synthesis method was ignored entirely. Based on the heating rate, the properties synthesized vary significantly. Those changes in properties occur both chemically and physically and significantly impact char reactions during gasification. Hence, kinetically comparing different fuels for any gasification process and understanding their char synthesis method is essential.
- Existence of maximum reaction rate: In most cases, the existence of a maximum rate was assumed at the beginning of the gasification reaction because of a high change in char properties. However, this assumption is not always valid. It may likely occur during CO_2 gasification, but when compared with H_2O (steam gasification), this observation may not be possible due to differences in the variation of properties.

Finally, it was suggested that the assumption char synthesis and maximum reaction rate assumptions are very implicit and need a proper understanding of scientific aspects [93].

3.6.2 Advances in Gasification Thermodynamics for Gasification Technology

Gasification technology is one promising technology for the thermochemical conversion of renewable sources such as waste and biomass. The advancement of thermodynamic considerations involves the development of advanced ways of interpreting gasification technology and processes in thermodynamic aspects. Here are some significant outlines of recent approaches to gasification technology.

- Solar-driven biomass gasification process: The thermodynamic study found a favorable approach to integrating cold gas efficiency and gas composition in parallel to the kinetic aspect without disturbing the energy balance of plant components [94].
- Biofuel production from gasification: An approach made to the thermodynamic equilibrium model for the gasification process in terms of concept comparison, process and heat integration, efficiency evaluation, and economic and ecological assessment has enabled the significance of thermodynamic studies [95].
- Assessment of supercritical water gasification: The thermodynamic analysis suggests a feasible approach for improving the energy and exergy efficiency analysis by enhancing some operating conditions, such as gasification temperature, slurry concentration, and heat transfer efficiency [96].
- Enhance biomass gasification efficiency: The thermodynamic equilibrium model based on a stoichiometric approach with modification based on empirical data shows reasonable accuracy in predicting gasification products. It was also noted that a modified thermodynamic equilibrium model could overcome the limitations that usually exist in the basic model [97].
- Biomass gasification for the production of hydrogen-rich syngas: Dynamic modeling and control studies have been conducted to optimize the approach's performance and productivity and capture the complex interactions between factors that influence gasification performance [98].

The overall approach suggests that understanding the thermodynamics of the thermochemical conversion process and its implications, particularly a gasification technology, is basically about a wide range from product to process efficiency, including the gasification kinetics and design of processes.

3.6.3 Understanding of the Roles of Transport Phenomena During Gasification Processes

The transport phenomena in gasification processes have a critical impact on overall gasification processes, from the design and operation of the front and back-end systems to the main gasification reactor [93, 99]. The role of transport phenomena varies concerning the design and operation of the reactor. For example, though different industrial gasifiers have certain limitations, the gasification operation associated with dynamic reactors and continuous operation of a gasification process shows a significant impact of transport phenomena on gasification processes—the major advantages of understanding the transport phenomena on its role in gasification technology [100, 101].

- Optimizing the process parameter can significantly improve the role of intraparticle and interparticle diffusion limitations.
- Understanding the solid-gas, solid-solid, and gas-gas interaction can be enlarged and implemented successfully by studying the correlation between heat, mass, and momentum transfer.
- Usefulness on the design of the gasification reactor by studying the hydrodynamic behavior in a gasification reactor.
- A relation between char gasification and the path of product gas formation can be easily presentable by understanding the effect of transport phenomena in a gasification reactor.
- Driven by the sustainability of gasification technology, some significant approaches can be taken in developing new gasification processes and reactors through advanced interpreting of the hydrodynamic of solid-gas interactions.

References

1. Jena, M. K., et al. (2021). Studies into the kinetic compensation effects of Loy Yang Brown coal during gasification in a steam environment—A mechanistic view. *Chemical Engineering Journal Advances, 8*, 100159.
2. Jena, M. K., et al. (2021). Mechanistic insights into the kinetic compensation effects during the gasification of Loy Yang Brown coal char in O2. *Industrial & Engineering Chemistry Research, 60*(49), 17881–17896.
3. Jena, M. K., et al. (2022). Steam gasification of low-rank coal chars: Insights into the kinetic compensation effects and physical significance of kinetic parameters. *Chemical Engineering Journal Advances, 11*, 100306.
4. Paiva, M., et al. (2021). Simulation of a downdraft gasifier for production of syngas from different biomass Feedstocks. *Chemical Engineering, 5*. https://doi.org/10.3390/chemengineering5020020
5. Havilah, P. R., et al. (2022). Biomass gasification in downdraft gasifiers: A technical review on production, up-gradation and application of synthesis gas. *Energies, 15*. https://doi.org/10.3390/en15113938
6. Khonde, R., Hedaoo, S., & Deshmukh, S. (2021). Prediction of product gas composition from biomass gasification by the method of Gibbs free energy minimization. *Energy Sources, Part A: Recovery, Utilization, and Environmental Effects, 43*(3), 371–380.
7. Frolov, S. M. (2021). Organic waste gasification: A selective review. *Fuels, 2*, 556–650. https://doi.org/10.3390/fuels2040033
8. Jena, M. K., Kumar, V., & Vuthaluru, H. (2022). Investigation into kinetic compensation effects for the production of hydrogen-rich gas during gasification of sub-bituminous coal char in varying gas environments. *International Journal of Hydrogen Energy, 47*(89), 37760–37773.

9. Akhtar, M. A., & Li, C.-Z. (2020). Mechanistic insights into the kinetic compensation effects during the gasification of biochar: Effects of the partial pressure of H_2O. *Fuel, 263*, 116632.

10. Akhtar, M. A., et al. (2018). Kinetic compensation effects in the chemical reaction-controlled regime and mass transfer-controlled regime during the gasification of biochar in O_2. *Fuel Processing Technology, 181*, 25–32.

11. Duman, G., Uddin, M. A., & Yanik, J. (2014). The effect of char properties on gasification reactivity. *Fuel Processing Technology, 118*, 75–81.

12. van Heek, K. H., & Mühlen, H. J. (1987). Effect of coal and char properties on gasification. *Fuel Processing Technology, 15*, 113–133.

13. Li, T., et al. (2017). Effects of gasification temperature and atmosphere on char structural evolution and AAEM retention during the gasification of Loy Yang brown coal. *Fuel Processing Technology, 159*, 48–54.

14. Akhtar, M. A., Zhang, S., & Li, C.-Z. (2019). Mechanistic insights into the kinetic compensation effects during the gasification of biochar in H_2O. *Fuel, 255*, 115839.

15. Li, T., et al. (2014). Effects of gasification atmosphere and temperature on char structural evolution during the gasification of collie sub-bituminous coal. *Fuel, 117*, 1190–1195.

16. Zhang, L., et al. (2015). Structural transformation of nascent char during the fast pyrolysis of mallee wood and low-rank coals. *Fuel Processing Technology, 138*, 390–396.

17. Wang, S., et al. (2015). Second-order Raman spectroscopy of char during gasification. *Fuel Processing Technology, 135*, 105–111.

18. Singh, S., et al. (2022). An integrated two-step process of reforming and adsorption using biochar for enhanced tar removal in syngas cleaning. *Fuel, 307*, 121935.

19. Li, C.-Z. (2004). *Advances in the science of Victorian brown coal*. Elsevier.

20. Feng, G., et al. (2019). Effect of O-containing functional groups and meso- and micropores on content and re-adsorption behavior of water in upgraded brown coal. *Fuel, 257*, 116100.

21. Feng, Y., et al. (2013). Surface modification of bituminous coal and its effects on methane adsorption. *Chinese Journal of Chemistry, 31*(8), 1102–1108.

22. Mengxia, Q., et al. (2018). Effects of CO_2/H_2O on the characteristics of chars prepared in $CO_2/H_2O/N_2$ atmospheres. *Fuel Processing Technology, 173*, 262–269.

23. Ye, C., et al. (2018). Characteristics of coal partial gasification experiments on a circulating fluidized bed reactor under CO2O2 atmosphere. *Applied Thermal Engineering, 130*, 814–821.

24. Xu, M., et al. (2018). A deep insight into carbon conversion during Zhundong coal molten salt gasification. *Fuel, 220*, 890–897.

25. Qing, M., et al. (2019). Relationships between structural features and reactivities of coal-chars prepared in CO_2 and H_2O atmospheres. *Fuel, 258*, 116087.

26. Zhao, D., et al. (2020). Effect of O-containing functional groups produced by preoxidation on Zhundong coal gasification. *Fuel Processing Technology, 206*, 106480.

27. Ban, Y., et al. (2022). Pyrolysis behaviors of model compounds with representative oxygen-containing functional groups in coal over calcium. *Fuel, 310*, 122247.

28. Tyler, R. J., & Schafer, H. N. S. (1980). Flash pyrolysis of coals: Influence of cations on the devolatilization behaviour of brown coals. *Fuel, 59*(7), 487–494.

29. Tyler, R. J. (1979). Flash pyrolysis of coals. 1. Devolatilization of a Victorian brown coal in a small fluidized-bed reactor. *Fuel, 58*(9), 680–686.

30. Li, C.-Z. (2007). Some recent advances in the understanding of the pyrolysis and gasification behaviour of Victorian brown coal. *Fuel, 86*(12), 1664–1683.

31. Li, C. Z., et al. (2000). Fates and roles of alkali and alkaline earth metals during the pyrolysis of a Victorian brown coal. *Fuel, 79*(3), 427–438.

32. Watanabe, W. S., & Zhang, D.-K. (2001). The effect of inherent and added inorganic matter on low-temperature oxidation reaction of coal. *Fuel Processing Technology, 74*(3), 145–160.

33. Lu, W., et al. (2020). Experimental study on the effect of preinhibition temperature on the spontaneous combustion of coal based on an MgCl2 solution. *Fuel, 265*, 117032.

34. Tang, Y. (2016). Inhibition of low-temperature oxidation of bituminous coal using a novel phase-transition aerosol. *Energy & Fuels, 30*(11), 9303–9309.
35. Quyn, D. M., Wu, H., & Li, C.-Z. (2002). Volatilisation and catalytic effects of alkali and alkaline earth metallic species during the pyrolysis and gasification of Victorian brown coal. Part I. Volatilisation of Na and Cl from a set of NaCl-loaded samples. *Fuel, 81*(2), 143–149.
36. Wu, H., et al. (2004). Volatilisation and catalytic effects of alkali and alkaline earth metallic species during the pyrolysis and gasification of Victorian brown coal. Part V. Combined effects of Na concentration and char structure on char reactivity. *Fuel, 83*(1), 23–30.
37. Wang, Y., et al. (2021). Effects of water-soluble sodium compounds on the microstructure and combustion performance of Shengli lignite. *ACS Omega, 6*(38), 24848–24858.
38. Bai, Y., et al. (2017). Coal char gasification in H_2O/CO_2: Release of alkali and alkaline earth metallic species and their effects on reactivity. *Applied Thermal Engineering, 112*, 156–163.
39. Du, C., Liu, L., & Qiu, P. (2017). Importance of volatile AAEM species to char reactivity during volatile–char interactions. *RSC Advances, 7*(17), 10397–10406.
40. Wu, L., et al. (2012). Effects of chemical forms of alkali and alkaline earth metallic species on the char ignition temperature of a Loy Yang coal under O_2/N_2 atmosphere. *Energy & Fuels, 26*(1), 112–117.
41. Li, T., et al. (2017). Effects of char chemical structure and AAEM retention in char during the gasification at 900°C on the changes in low-temperature char-O_2 reactivity for collie sub-bituminous coal. *Fuel, 195*, 253–259.
42. Kim, H.-S., et al. (2013). Fluidized bed drying of Loy Yang brown coal with variation of temperature, relative humidity, fluidization velocity and formulation of its drying rate. *Fuel, 105*, 415–424.
43. Rao, Z., et al. (2015). Recent developments in drying and dewatering for low rank coals. *Progress in Energy and Combustion Science, 46*, 1–11.
44. Ullah, H., et al. (2018). Hydrothermal dewatering of low-rank coals: Influence on the properties and combustion characteristics of the solid products. *Energy, 158*, 1192–1203.
45. Dang, H., et al. (2021). Study on chemical bond dissociation and the removal of oxygen-containing functional groups of low-rank coal during hydrothermal carbonization: DFT calculations. *ACS Omega, 6*(39), 25772–25781.
46. Guo, H., et al. (2018). Effect of moisture on the desorption and unsteady-state diffusion properties of gas in low-rank coal. *Journal of Natural Gas Science and Engineering, 57*, 45–51.
47. Chen, M.-Y., et al. (2018). Impact of inherent moisture on the methane adsorption characteristics of coals with various degrees of metamorphism. *Journal of Natural Gas Science and Engineering, 55*, 312–320.
48. Zhang, W., et al. (2018). Molecular dynamics simulations of interaction between sub-bituminous coal and water. *Molecular Simulation, 44*(9), 769–773.
49. Kershaw, J. R., et al. (2000). Fluorescence spectroscopic analysis of tars from the pyrolysis of a Victorian Brown coal in a wire-mesh reactor. *Energy & Fuels, 14*(2), 476–482.
50. Sathe, C., Pang, Y., & Li, C.-Z. (1999). Effects of heating rate and ion-exchangeable cations on the pyrolysis yields from a Victorian Brown coal. *Energy & Fuels, 13*(3), 748–755.
51. Song, H., Liu, G., & Wu, J. (2016). Pyrolysis characteristics and kinetics of low rank coals by distributed activation energy model. *Energy Conversion and Management, 126*, 1037–1046.
52. Dwivedi, K. K., et al. (2019). Pyrolysis characteristics and kinetics of Indian low rank coal using thermogravimetric analysis. *International Journal of Coal Science & Technology, 6*(1), 102–112.
53. Kang, T.-J., et al. (2017). Comparison of catalytic pyrolysis and gasification of Indonesian low rank coals using lab-scale bubble fluidized-bed reactor. *Korean Journal of Chemical Engineering, 34*(4), 1238–1249.
54. Wornat, M. J., & Sakurovs, R. (1996). Proton magnetic resonance thermal analysis of a brown coal: Effects of ion-exchanged metals. *Fuel, 75*(7), 867–871.

55. Quyn, D. M., et al. (2002). Volatilisation and catalytic effects of alkali and alkaline earth metallic species during the pyrolysis and gasification of Victorian brown coal. Part II. Effects of chemical form and valence. *Fuel, 81*(2), 151–158.

56. Sonoyama, N., et al. (2006). Interparticle desorption and re-adsorption of alkali and alkaline earth metallic species within a bed of Pyrolyzing char from pulverized Woody biomass. *Energy & Fuels, 20*(3), 1294–1297.

57. Wood, B. J., & Sancier, K. M. (1984). The mechanism of the catalytic gasification of coal char: A critical review. *Catalysis Reviews, 26*(2), 233–279.

58. Lang, R. J. (1986). Anion effects in alkali-catalysed steam gasification. *Fuel, 65*(10), 1324–1329.

59. Suzuki, T., Ohme, H., & Watanabe, Y. (1992). A mechanism of sodium-catalyzed carbon dioxide gasification of carbon investigation by pulse and TPD techniques. *Energy & Fuels, 6*(4), 336–343.

60. Suzuki, T., Ohme, H., & Watanabe, Y. (1992). Alkali metal catalyzed carbon dioxide gasification of carbon. *Energy & Fuels, 6*(4), 343–351.

61. Chen, S. G., & Yang, R. T. (1992). Mechanism of alkali and alkaline earth catalyzed gasification of graphite by CO_2 and H_2O studied by electron microscopy. *Journal of Catalysis, 138*(1), 12–23.

62. Quyn, D. M., et al. (2003). Volatilisation and catalytic effects of alkali and alkaline earth metallic species during the pyrolysis and gasification of Victorian brown coal. Part IV. Catalytic effects of NaCl and ion-exchangeable Na in coal on char reactivity☆. *Fuel, 82*(5), 587–593.

63. Meijer, R., et al. (1991). Catalyst loss and retention during alkali-catalysed carbon gasification in CO_2. *Carbon, 29*(7), 929–941.

64. Takarada, T., Tamai, Y., & Tomita, A. (1985). Reactivities of 34 coals under steam gasification. *Fuel, 64*(10), 1438–1442.

65. Quyn, D. M., Hayashi, J.-I., & Li, C.-Z. (2005). Volatilisation of alkali and alkaline earth metallic species during the gasification of a Victorian brown coal in CO_2. *Fuel Processing Technology, 86*(12), 1241–1251

66. Li, X., Hayashi, J.-I., & Li, C.-Z. (2006). FT-Raman spectroscopic study of the evolution of char structure during the pyrolysis of a Victorian brown coal. *Fuel, 85*(12), 1700–1707.

67. Tay, H.-L., et al. (2014). A preliminary Raman spectroscopic perspective for the roles of catalysts during char gasification. *Fuel, 121*, 165–172.

68. Marcantonio, V., et al. (2023). Modeling of biomass gasification: From thermodynamics to process simulations. *Energies, 16*. https://doi.org/10.3390/en16207042

69. Ajorloo, M., et al. (2022). Recent advances in thermodynamic analysis of biomass gasification: A review on numerical modelling and simulation. *Journal of the Energy Institute, 102*, 395–419.

70. Rupesh, S., Muraleedharan, C., & Arun, P. (2016). ASPEN plus modelling of air–steam gasification of biomass with sorbent enabled CO_2 capture. *Resource-Efficient Technologies, 2*(2), 94–103.

71. Rokni, M. (2015). Thermodynamic analyses of municipal solid waste gasification plant integrated with solid oxide fuel cell and Stirling hybrid system. *International Journal of Hydrogen Energy, 40*(24), 7855–7869.

72. Ramos, A., Monteiro, E., & Rouboa, A. (2019). Numerical approaches and comprehensive models for gasification process: A review. *Renewable and Sustainable Energy Reviews, 110*, 188–206.

73. Mazaheri, N., et al. (2019). Systematic review of research guidelines for numerical simulation of biomass gasification for bioenergy production. *Energy Conversion and Management, 183*, 671–688.

74. Ahmed, T. Y., et al. (2012). Mathematical and computational approaches for design of biomass gasification for hydrogen production: A review. *Renewable and Sustainable Energy Reviews, 16*(4), 2304–2315.

75. Puig-Arnavat, M., Bruno, J. C., & Coronas, A. (2012). Modified thermodynamic equilibrium model for biomass gasification: A study of the influence of operating conditions. *Energy & Fuels, 26*(2), 1385–1394.
76. Patra, T. K., & Sheth, P. N. (2015). Biomass gasification models for downdraft gasifier: A state-of-the-art review. *Renewable and Sustainable Energy Reviews, 50*, 583–593.
77. Puig-Arnavat, M., Bruno, J. C., & Coronas, A. (2010). Review and analysis of biomass gasification models. *Renewable and Sustainable Energy Reviews, 14*(9), 2841–2851.
78. Násner, A. M. L., et al. (2017). Refuse Derived Fuel (RDF) production and gasification in a pilot plant integrated with an Otto cycle ICE through Aspen plus™ modelling: Thermodynamic and economic viability. *Waste Management, 69*, 187–201.
79. Renkel, M. F., & Lümmen, N. (2018). Supplying hydrogen vehicles and ferries in Western Norway with locally produced hydrogen from municipal solid waste. *International Journal of Hydrogen Energy, 43*(5), 2585–2600.
80. Couto, N. D., Silva, V. B., & Rouboa, A. (2016). Thermodynamic evaluation of Portuguese municipal solid waste gasification. *Journal of Cleaner Production, 139*, 622–635.
81. Loha, C., et al. (2014). Advances in mathematical modeling of fluidized bed gasification. *Renewable and Sustainable Energy Reviews, 40*, 688–715.
82. Bermudez, J. M., & Fidalgo, B. (2016). Chapter 15 – Production of bio-syngas and bio-hydrogen via gasification. In R. Luque et al. (Eds.), *Handbook of biofuels production* (2nd ed., pp. 431–494). Woodhead Publishing.
83. Francia, V., Wu, K., & Coppens, M.-O. (2021). Dynamically structured fluidization: Oscillating the gas flow and other opportunities to intensify gas-solid fluidized bed operation. *Chemical Engineering and Processing – Process Intensification, 159*, 108143.
84. Lipiński, W., et al. (2021). Progress in heat transfer research for high-temperature solar thermal applications. *Applied Thermal Engineering, 184*, 116137.
85. von Berg, L., et al. (2022). Multi-scale modelling of fluidized bed biomass gasification using a 1D particle model coupled to CFD. *Fuel, 324*, 124677.
86. Siedlecki, M., De Jong, W., & Adrian, H. M. V. (2011). Fluidized bed gasification as a mature and reliable Technology for the Production of bio-syngas and applied in the production of liquid transportation fuels—A review. *Energies, 4*, 389–434. https://doi.org/10.3390/en4030389
87. Phounglamcheik, A., et al. (2022). The significance of intraparticle and interparticle diffusion during CO2 gasification of biomass char in a packed bed. *Fuel, 310*, 122302.
88. Gómez-Barea, A., & Leckner, B. (2010). Modeling of biomass gasification in fluidized bed. *Progress in Energy and Combustion Science, 36*(4), 444–509.
89. Israelsson, M., Larsson, A., & Thunman, H. (2014). Online measurement of elemental yields, oxygen transport, condensable compounds, and heating values in gasification systems. *Energy & Fuels, 28*(9), 5892–5901.
90. Zagorščak, R., et al. (2019). Underground coal gasification—A numerical approach to study the formation of syngas and its reactive transport in the surrounding strata. *Fuel, 253*, 349–360.
91. Yan, M., Afxentiou, N., & Fokaides, P. A. (2021). The state of the art overview of the biomass gasification technology. *Current Sustainable/Renewable Energy Reports, 8*(4), 282–295.
92. Moreira, R., et al. (2021). Clean syngas production by gasification of lignocellulosic char: State of the art and future prospects. *Journal of Industrial and Engineering Chemistry, 101*, 1–20.
93. Mahinpey, N., & Gomez, A. (2016). Review of gasification fundamentals and new findings: Reactors, feedstock, and kinetic studies. *Chemical Engineering Science, 148*, 14–31.
94. Freda, C., et al. (2022). Thermodynamic improvement of solar driven gasification compared to conventional one. *Energy, 261*, 124953.
95. Sikarwar, V. S., et al. (2017). Progress in biofuel production from gasification. *Progress in Energy and Combustion Science, 61*, 189–248.

96. Chen, J., et al. (2020). Assessment of supercritical water gasification process for combustible gas production from thermodynamic, environmental and techno-economic perspectives: A review. *Energy Conversion and Management, 226*, 113497.

97. Silva, I. P., et al. (2019). Thermodynamic equilibrium model based on stoichiometric method for biomass gasification: A review of model modifications. *Renewable and Sustainable Energy Reviews, 114*, 109305.

98. Hussain, M., et al. (2023). Recent advances in dynamic modeling and control studies of biomass gasification for production of hydrogen rich syngas. *RSC Advances, 13*(34), 23796–23811.

99. Perkins, G., & Sahajwalla, V. (2007). Modelling of heat and mass transport phenomena and chemical reaction in underground coal gasification. *Chemical Engineering Research and Design, 85*(3), 329–343.

100. Janajreh, I., et al. (2021). A review of recent developments and future prospects in gasification systems and their modeling. *Renewable and Sustainable Energy Reviews, 138*, 110505.

101. Di Renzo, A., Scala, F., & Heinrich, S. (2021). Recent advances in fluidized bed hydrodynamics and transport phenomena—Progress and understanding. *PRO, 9*. https://doi.org/10.3390/pr9040639

102. Manoj, K., Jena, V. K., & Hari V. Usefulness of insights into the kinetic compensation effects in the kinetic analysis of the coal gasification process. *The Canadian Journal of Chemical Engineering*. https://doi.org/10.1002/cjce.25477

Chapter 4
Gasification Reactor and Processes

The Heart of Energy Transformation: From Waste to Value-Added Products

4.1 Introduction

The gasification process partially oxidizes carbonaceous material in a limited oxygen environment. The design of a chemical reactor on an industrial scale is primarily focused on intrinsic kinetic parameters. However, to develop true kinetic parameters for gasification, performing lab experiments in a specified reactor would enable us to achieve this objective. Moreover, from the lab-scale gasification process, the rate can be defined from the intrinsic kinetic parameter derived from the experimented rate law value. As the gasification process is a thermochemical conversion of solid carbonaceous materials, the proper execution of the gasification reaction, which means a solid-gas heterogeneous reaction and an appropriate theoretical and practical understanding of the process and hydrodynamics of the reactor, is essential [1, 2]. This chapter aims to provide a comprehensive understanding of the gasification reactor process. It will discuss how the reactor and processes are interrelated and the significant impact of the gasification reactor on overall performance. Some of the speculated conclusions will be drawn by comparing and contrasting different reactors, guiding you through the complexities of this field.

M. K. Jena, H. B. Vuthaluru, *Gasification Technology*,
https://doi.org/10.1007/978-3-031-71044-5_4

4.2 Classification of Modern Gasification Reactor

The modern gasification reactor is a complex system that requires careful consideration. It is classified into four major types: a wire-mesh reactor, thermogravimetric analyzer, one-stage fluidized-bed/fixed-bed reactor, and two-stage fluidized-bed/fixed-bed reactor. This classification is based on the weight of the sampler each reactor can handle and the prime objective of performing gasification in that particular reactor. For instance, the wire-mesh reactor and thermogravimetric analyzer can handle samples in the milligrams (mg) range. In contrast, in pilot scale studies, a one-stage/two-stage fluidized bed reactor can handle samples in the gram (gm) range on a lab scale and kilogram (kg) range. Therefore, the choice of reactor is determined by the amount of feedstock it can handle, which is based on the prime objective of the gasification process [3, 4]. In general, a reactor that can handle milligrams is primarily used for fundamental studies to understand the process and the feedstock. Reactors that can have capacities in grams or kilograms are primarily subjected to scaleup studies to be semi-pilot and/or plant scale. The following section will provide a brief discussion of each reactor.

4.2.1 Wire-Mesh Reactor (WMR)

A WMR is usually intended to be used to understand the pyrolysis and gasification in a micro-level mechanism [5, 6]. This reactor can handle any solid fuel, coal, biomass, and other solid carbon fuel materials. The major benefit of this WMR is that it can be operated at temperatures of around ~1000 °C at a very high heating rate of up to 1000 °C/min. A picture and schematic view of the experimental setup aligned with WMR are shown in Figs. 4.1a and b [7, 8]. The reactor mainly consists of a temperature measuring segment, a control system, and a quartz chamber where thermochemical operation is performed. A set of thermocouples was used to measure the lateral variation of temperature to be consistent with the desired temperature. A computer controls heating and temperature measurements. Based on the design of the experimental matrix, the inert gas and gasifying agent are fed to the reactor accordingly. The inert gas could be helium (He), nitrogen (N_2) and argon (Ar) for executing pyrolysis operation. The gasifying agents are carbon dioxide (CO_2), oxygen (O_2), and a mixture of both based on the requirements and conditions. In briefly explaining the use of WMR, before every experiment, WMR was punched in a specific condition to form a circular reaction zone. After that, the required sample was distributed to the reactor. A steel mesh was folded on the reacting zone to avoid entraining the sample from the WMR during operation. The WMR reactor is specially designed to understand the reaction mechanism; hence, in some instances, there would be some difficulties under the hydrodynamic mechanism inside the reactor. However, in some cases, plug flow condition was assumed to ascertain some conclusions.

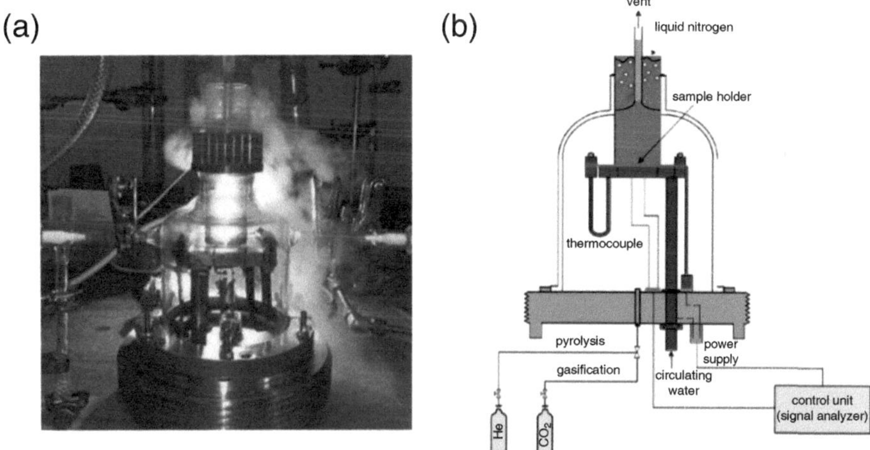

Fig. 4.1 Picture of wire mesh reactor (**a**) and schematic view of the wire mesh reactor system (**b**) [7, 8]

4.2.2   Thermogravimetric Analyzer

Like WMR, the thermogravimetric analyzer (TGA) can handle some milligrams for any solid feedstock. Based on technology, the thermogravimetric analyzer can be both automated and manual. A thermogravimetric analyzer can be used for pyrolysis, gasification, and some other kinds of characterization, i.e., proximate analysis, calorific value, and heat of pyrolysis. The operation of TGA is very simple if the aim of experimenting is predefined [9, 10]. The sample containing the crucible needs to measure the empty weight before performing the TGA experiment. Then, the feed sample was placed with variation in weight, and the product gas composition was measured in time. A TGA experiment can be performed to a maximum temperature of 900 °C to 1000 °C, and the heating rate can be adjusted from 5 °C/min to 50 °C/min or even more based on the type and model of TGA equipment. The experiment can be designed by keeping the prime objective, and findings are made to understand the gasification kinetics most of the time. The preferable gases that can be used for executing pyrolysis, gasification, and combustion are defined according to their intended reaction mechanism (a schematic view of the TGA and TGA analysis system, including the experimental setup, is shown in Fig. 4.2a–b) [11–13]. Similar to the WMR reactor, it operated in batch mode, with basic differences in heater rate. In the case of WMR, a very high heating rate gasification reaction can be performed compared to TGA at the lab scale level.

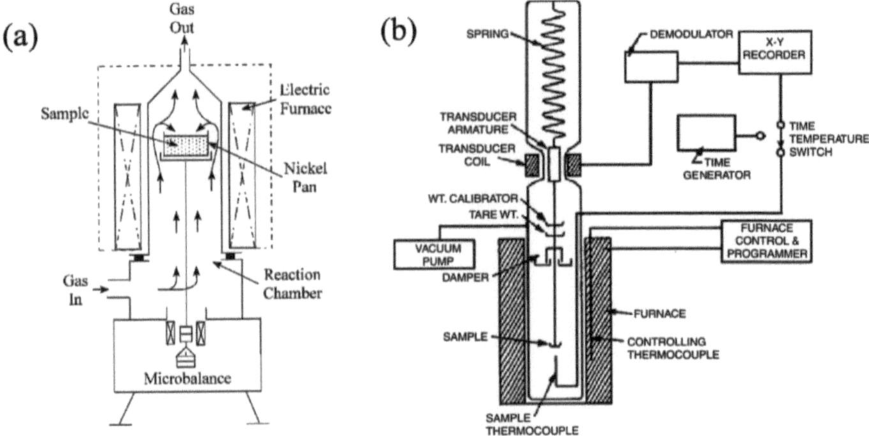

Fig. 4.2 Schematic diagram of the thermogravimetric analyzer (**a**) and thermogravimetric analysis system (**b**) [11–13]

4.2.3   One-Stage Fluidized-Bed/Fixed-Bed Reactor

The one-stage fluidized bed /fixed-bed reactor is an industrial reactor where we can analyze hydrodynamic and reaction mechanisms for any gasification operation in different gasifying environments. As shown in Fig. 4.3a, the schematic view of the reactor looks a bit complex compared to the WMR and TGA reactors. The reaction is designed to insert multiple thermocouples inside the reactor to measure the pyrolysis/gasification operation's temperature uniformly. It is operated in batch mode with slow and fast feeding processes. The fluidizing gas and gasifying agent are fed from the bottom into the reactor/reacting zone through a flit or distributor. During a continuous operation of gasification processes, the fuel particle could be entrained from the bed due to a change in density or particle size. To avoid this situation, a flit/ filter was set at the exit of the reactor to prevent the entrainment and act as a fixed bed mechanism over the flit. Therefore, the reactor and mechanism were defined as One-stage fluidized-bed/fixed-bed reactor. Apart from investigating hydrodynamic analysis, the other advantage of this reactor is that it can be scaled up from a lab scale to an industrial scale reactor for performing large-scale operations utilizing a wide range of feedstock. Also, it provides another advantage: performing fundamental investigations on gasification mechanisms to get a broader view of applying a different feed into a generic reactor. The schematic view of the experimental setup is shown in Fig. 4.3b as can be. An online gas measuring system may be required to monitor product composition to understand the reaction mechanism continuously.

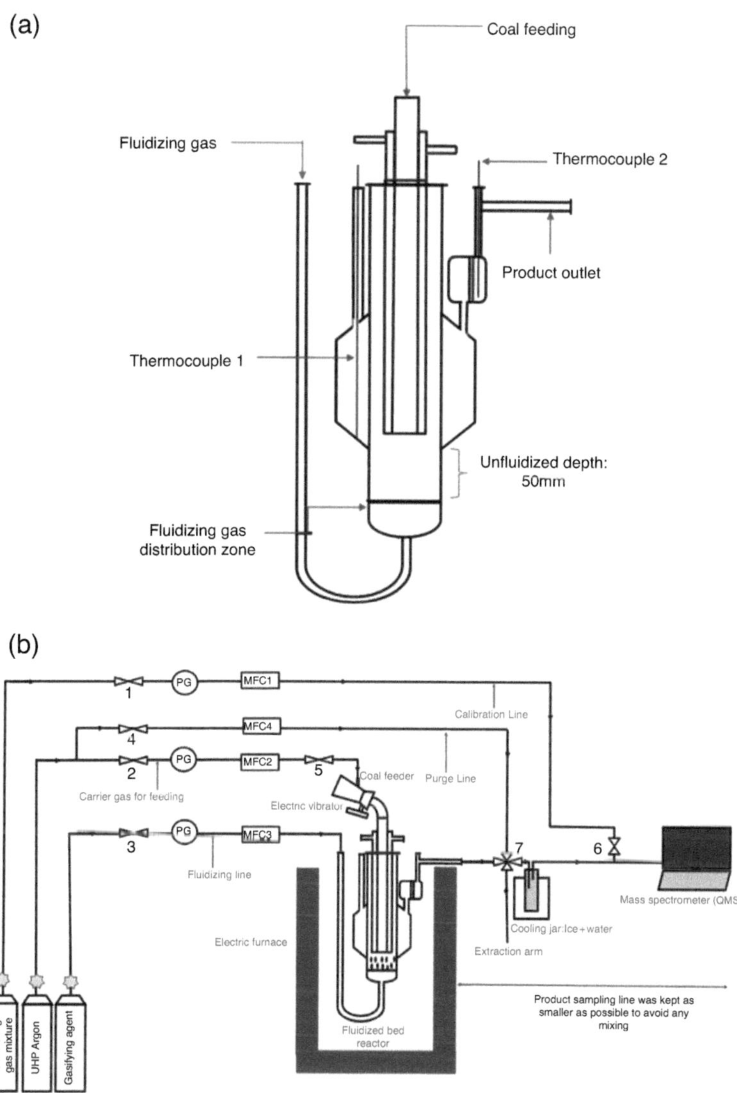

Fig. 4.3 A schematic diagram of the fluidized bed reactor (**a**) simplified from Asadullah et al. [14] and a schematic outline of the experimental setup used for this study (**b**) [15]

4.2.4   Two-Stage Fluidized-Bed/Fixed-Bed Reactor

Usually, a three flit, two-stage fluidized-bed/fixed-bed reactor was primarily used for catalytic steam reforming of tar (Fig. 4.4a) [16]. That lab-scale reactor operates slightly differently from a one-stage fluidized-bed/fixed-bed reactor. To briefly explain this reactor's operation, steam reforming of tar using biomass biochar as a

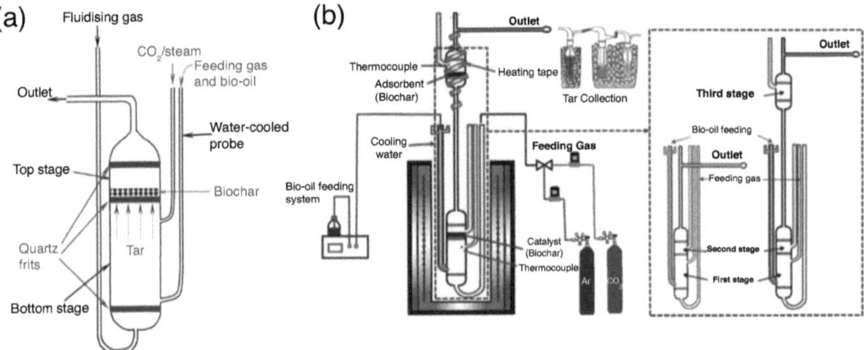

Fig. 4.4 Schematic view of a two-stage fluidized bed (**a**) and its experimental setup (**b**) [16, 17]

catalyst was used for explaining here. The biochar, considered a catalyst, was activated using the same reactor before the primary reforming reaction. The reactor is made of quartz, so it maintains its thermal operation at around 1100 °C for longer. At each lab-scale operation, the reactor was taken out from the electric furnace to stop the further occurrence of reforming reactions. Additionally, as shown in Fig. 4.4b, a tar collection arrangement was made to receive the effluent product for analysis during lab scale analysis. However, there are some constraints to using this kind of setup for industrial-scale analysis, such as condensation of tar, collection of product gas, and activation of biomass biochar in large mode. In addition, built-in back pressure could be possible when safety issues are encountered. However, understanding the process mechanism using this kind of reactor in a scale setup (as presented in Fig. 4.4b) [17] would provide a platform for prospects to think about developing the technology commercially.

4.3 Trends of Gasification Processes: Based on the Role of Pyrolysis

The type of gasification process set the trend of the gasification process. Mechanistically, the pyrolysis and gasification of char/biochar co-occur during the initial stage of thermal processing of the feedstock. However, if char gasification is desired for the processes, forming char/biochar from the raw sample through pyrolysis is the primary step. The process was segregated into two types during pyrolysis based on the heating rate. One is slow pyrolysis, where the heating rate of the sample is very low and leads to the slow release of volatile matter. Another aspect is that the fast pyrolysis mechanism occurs upon heating the sample in an inert atmosphere at a high heating rate. The product char sample obtained from fast and slow pyrolysis shows different properties and could show different pathways for the respective char gasification [18–20]. A brief discussion on the slow and fast pyrolysis mechanism and the subsequent char gasification will be discussed in the following section.

4.3.1   Fast Pyrolysis

The slow pyrolysis studies were usually conducted in a wire mesh reactor. The study has been done on fast pyrolysis (heating rate of 1000 °K/s) of low-rank fuel, including biomass (Collie-sub bituminous coal, Loy Yang brown coal, and Australian mallee wood biomass), shows a significant change in char properties with variation in holding time ranging from 0 s to 50 s. The result of a change in char properties of different feedstock indicates that a significant change in ring condensation occurs with the change in pyrolysis temperature (Fig. 4.5) [21]. With the increased holding time, the ring condensation has primarily occurred for those three low-rank fuels.

Furthermore, the gasification study of char (produced through fast pyrolysis) in a CO_2 environment shows that the holding of char (specially Loy Yang brown coal and Collie-sub-bituminous coal) in 1000 °C and 1200 °C for specific periods has a more significant impact on the rapid formation of O-containing functional group on char surface with selective consumption of smaller aromatic ring (Figs. 4.6 and 4.7). In addition, no significant change in the path of char gasification was noticed between the operational temperature of 1000 °C to 1200 °C for both coal samples [22].

This indicates that char derived from fast pyrolysis of low-rank fuel at different temperatures has no significant impact on subsequent gasification processes. If the gasification process has occurred at higher temperatures in a CO_2 environment, the process of char gasification is broadly the same for low-rank fuels. However, there are certain benefits of optimizing the process from the aspect of fast pyrolysis, as listed below.

- The difference in nascent char properties obtained from pyrolysis (different fuels and temperatures) will allow a platform for optimizing the char gasification reactivity.
- The fast pyrolysis reaction usually occurs very quickly, allowing the gasification process to happen very soon.

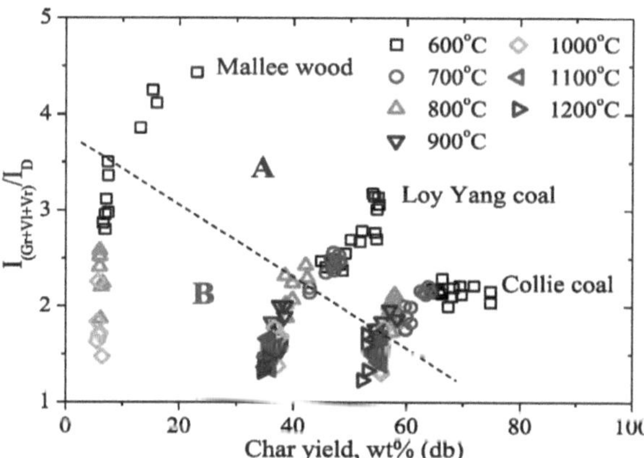

Fig. 4.5 Change in char properties with change in pyrolysis temperature during fast pyrolysis of low-rank fuel [21]

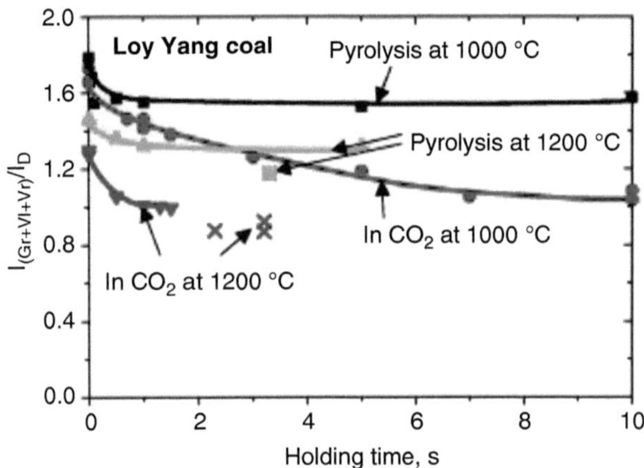

Fig. 4.6 Changes in $I_{(Gr + Vl + Vr)}/I_D$ as a function of holding time at 1000 °C and 1200 °C during the pyrolysis in He or the gasification in CO_2 of Loy Yang coal [22]

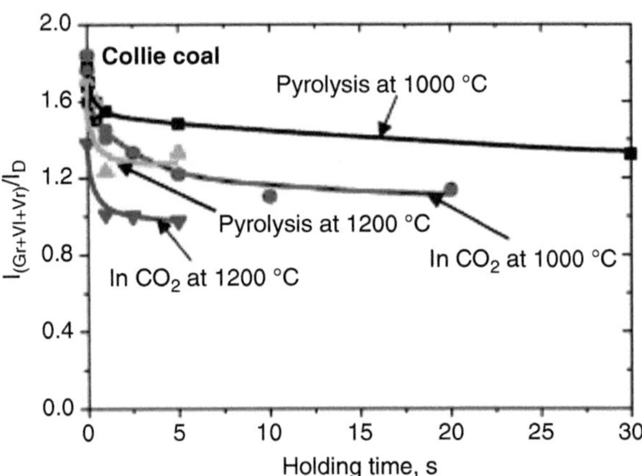

Fig. 4.7 Changes in $I_{(Gr + Vl + Vr)}/I_D$ as a function of holding time at 1000 °C and 1200 °C during the pyrolysis in He or the gasification in CO_2 for Collie coal [22]

In addition to the above true benefit, the significant challenges associated with fast pyrolysis are listed below.

- The usual method has too little response time; a high-sensitivity control system is needed for effectiveness in a pilot and/or plant scale.
- As the thermal operation occurs very fast, in the case of bigger particles, possibilities of intra-particle mass and heat transfer occur, resulting in observing and noticing the actual temperature inside the particle.
- As the solid fuel sample reacts in a very hot environment, some possibilities exist to influence the temperature measurement (radiation, gas convection, etc.).
- Overall, scaling up gasification technology associated with fast pyrolysis operations is very complicated, as fast pyrolysis operations are suitable for lab scale with very low feed rates.

4.3.2   Slow Pyrolysis

The slow pyrolysis operation is carried out with a very low heating rate compared to the fast pyrolysis operation, which is about 0.01 °C/min to 80 °C/min [23, 24]. The slow pyrolysis process maximizes char production compared to fast pyrolysis. Based on the mechanism process and intended to receive the required product, there are two crucial types of slow pyrolysis: carbonization (heating for days: more extended period) and conventional pyrolysis (heating for 5–30 min). In the carbonization process, the primary focus was producing more char while ignoring the vapor products. However, subjecting to the gasification process, which is a subsequence of pyrolysis, the conventional slow pyrolysis is preferred, elaborated briefly in the following sections. Conventional pyrolysis is mainly adopted in industries that allow the collection of char, oil, and non-condensable vapors. A study has shown the properties and behavior of agri waste and food residue products, and it shows that maximized biochar as product residue was obtained at 600 °C, at a heating rate of 5 °C/min for a holding time of 60 min [25]. However, the enhancement of biochar production with optimized process conditions is limited to the kind of feedstock properties, which include density, particle size, moisture, volatile matter, and fixed carbon content of the sample. The characteristics of char (reactivity of char) derived from pyrolysis depend on heating rate, pyrolysis temperature, and feedstock type. The derived char from slow pyrolysis can present low reactivity because of low catalytic inorganic content compared to chard derived from the fast pyrolysis method. Contrary to this, a low heating rate during fast pyrolysis results in the formation of high surface areas in contrast to a high heating rate (Fig. 4.8). Therefore, the porous nature of char enhances the reactivity by compensating the decremental effect of low inorganic catalytic content in the case of some feedstocks, e.g., agri waste, forestry residue, wood chips, etc. [26].

Consider the recent trend of slow pyrolysis, which mainly produces gaseous products and char. A slow pyrolysis operation is usually carried out in a temperature range of 300 °C to 550 °C. The vapor is usually not condensed during the process but contributes to heat. Therefore, torrefaction is a form of slow pyrolysis that occurs in a temperature range of 225 °C to 300 °C [26]. Significant benefits are outlined while operating a thermal system through slow pyrolysis operation for energy and char/biochar production.

- Simple equipment, ease of operation, and control in a large-scale operation.
- A wide range of feedstock variations in physical and chemical properties can be handled efficiently.
- Slow pyrolysis technology is one kind of standard technology and might require additional energy compared to fast pyrolysis.
- Biochar production can be enhanced per desired conditions by efficiently optimizing the operating conditions.

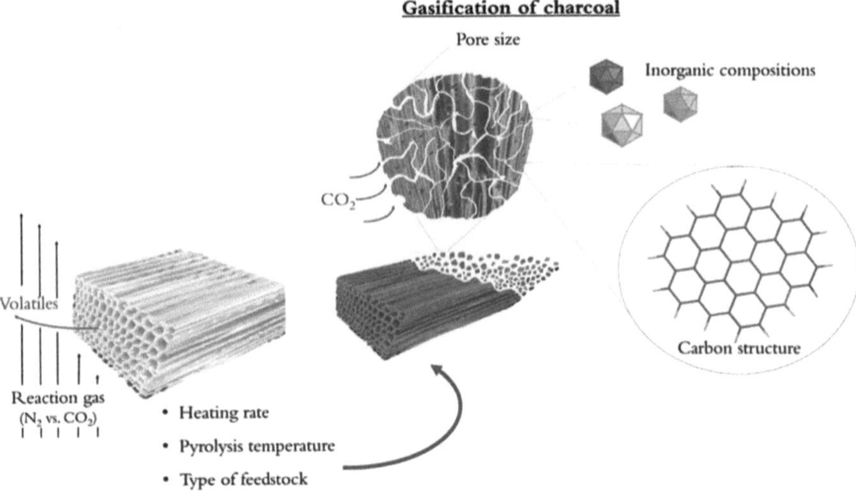

Fig. 4.8 Schematic view on effects of pyrolysis conditions and feedstocks on the properties and gasification reactivity of charcoal derived from woodchips [26]

4.4   Model of Reactor and Gasification Systems

The models of reactor and gasification systems are essential, as this will provide a platform to understand the process, reaction mechanism, particle behavior, and other interactions profoundly and thoroughly [3, 4]. Also, in many commercial approaches, forming a prototype via the design of a model of the reactor or process would enable the refine and optimize the process parameters for large-scale/ industrial-scale operations. The modeling approach is defined by keeping the basis of requirement and will characterize the primary requirement from the gasification operation. Hence, the reason for the modeling aspect of the gasifying reactor is that an approach will be given to the model of the particular reactor, focusing on the hydrodynamics of the reactor under hot and cold flow conditions. The gasification process model will provide a platform to design a process model that includes the front and back of a gasifying system. The process model will also allow the optimization of the process parameters to achieve a safer operation at a gasification plant. Generally, a process model is the model of a gasification plant, including all required operations to be carried out upstream and downstream of a gasifying reactor, and it also provides knowledge on balancing energy and mass of a process. The hydrodynamic model is only specific to gasifying reactors, providing knowledge on the mechanism of heat, mass, and momentum transfer between the particle-particle and particle-gasifying media with a specified reactor. The following section will briefly discuss recent advances in the modeling approach.

4.4.1 Process Model of the Different Gasifying Units

Various process models are defined for different gasifying units. Some are process models of wood gasification units combined with heat and power plants [27]., a process model of Lurgi fixed bed coal gasifier in a synthetic natural gas (SNG) plant [12], and a process model of a two-stage thermal process for a waste gasification plant [28], etc. Those defined process models and others are based on the thermodynamic approach. The configuration of the plant is illustrated in Fig. 4.9 (as an example); the overall process of the plant includes the drying of wood chips that initially occurred and then supplied to a dual bed fluidizer (DBF). In particular, for the process model of a wood gasification unit combined with a heat and power plant, the gasifier is a dual fluidized bed with three sections. Those three sections describe wood pyrolysis, secondary reactions, and char combustions. The flue gas exit from the combustor compartment was cleaned and cooled to meet the internal combustion (IC) gas engine requirements.

The process model developed using the tool Aspen Plus (Fig. 4.10) would also benefit from globally measuring mass and energy balance throughout the processes to optimize the process parameters better before scaling up the process to an industrial/plant level. The illustrated picture was presented here as an example to explain things in more detail. When operating the Aspen Plus-defined model simulation, the global input and output are assigned to the model when defining each sub-section of

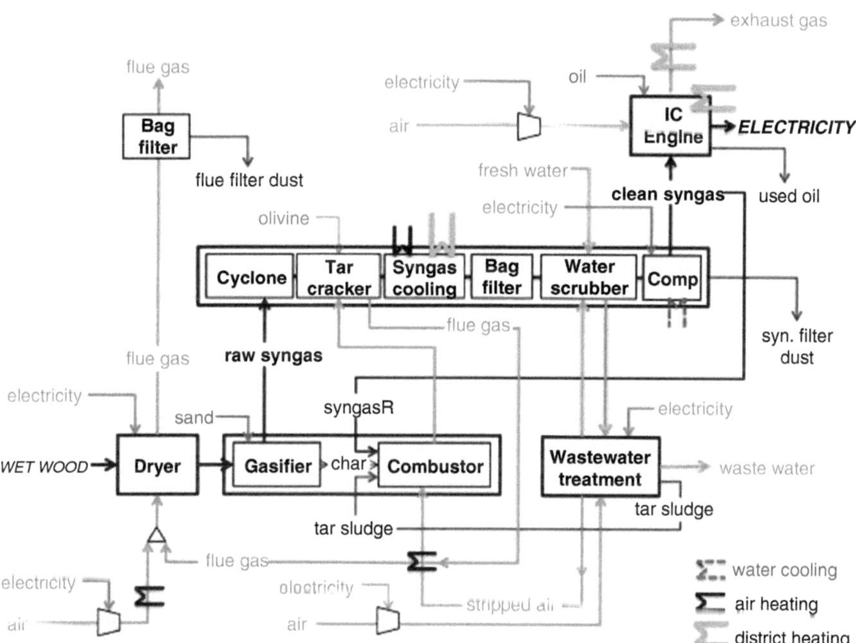

Fig. 4.9 Schematic structure of the gasification unit model [27]

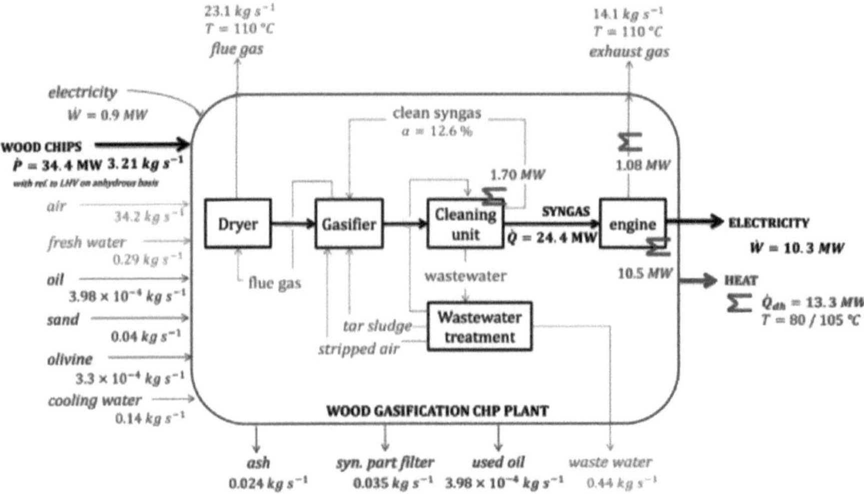

Fig. 4.10 Global mass and energy balance of the wood gasification combined heat and power (CHP) plant, obtained from the Aspen Plus® simulation [27]

the model. Finally, the result obtained from the simulation of the designed process model is validated by other available experimented literature and industrial data.

Moreover, the other process model uses a gasifying reactor as the primary unit. The Aspen Plus, simulation calculation procedure, plays a dominant role in measuring the value of process parameters. For example, in the case of the process model of the Lurgi fixed-bed coal gasifier in an SNG plant, the coal was treated as a mixture of elementary substances such as C(s), H, N, Cl, S(s), O and ash because those elements are going to describe the chemical formula of coal. There is no predefined chemical formula available for the coal. According to the actual process of pressurized Lurgi gasifier, the process model is usually defined by five different Aspen blocks to simulate the overall process. This includes the gasifying drying zone, pyrolysis zone, gasification zone, combustion zone, and overall heat recovery unit (as shown in Fig. 4.11) [29]. In conclusion, analyzing a gasification system through the design of a process model has broader implications for understanding the process parameters of one subsystem in relation to another subsystem. This model approach will provide knowledge-based operation of a gasification plant in a safer condition, resulting in the flexibility of the operating conditions as per the desired situation.

4.4.2 Hydrodynamic Analysis of the Gasifying Reactors

A hydrodynamic modeling analysis of the reactor is specially intended to observe the dynamics of the reactor in steady and unsteady conditions, which are usually very difficult to observe in a regular industrial-scale operation. Some hydrodynamic models are developed to observe the heat, mass, and momentum profile through commercially available tools such as computational fluid dynamics (CFD), where

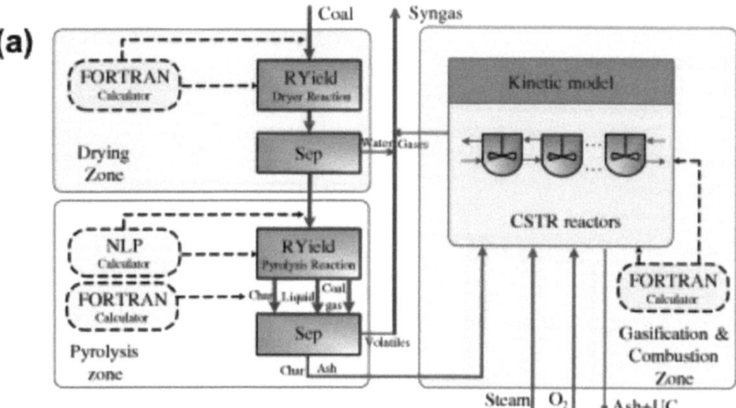

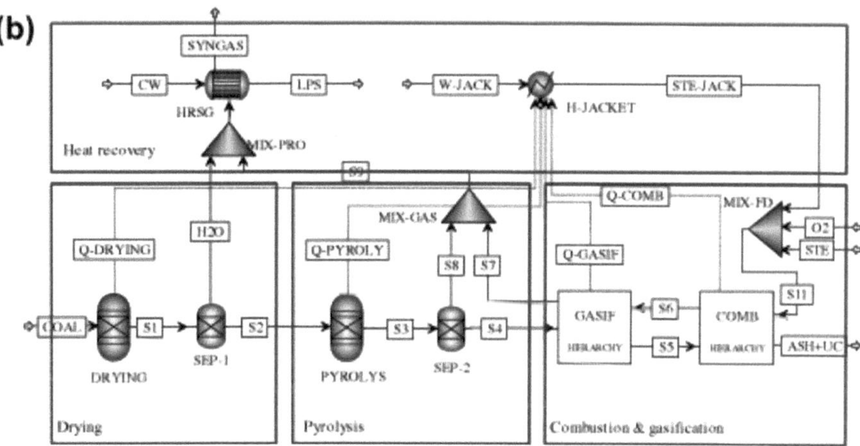

Fig. 4.11 Schematic view of Aspen Plus simulation calculation procedure (**a**) and Aspen Plus simulation model of pressurized Lurgi gasifier (**b**) [29]

some defined inbuild mathematical models are available as process model tool Aspen Plus where a set inbuild thermal model for different reactor and sub-processes are available. Mathematical models are sometimes developed from heat, mass, and momentum balance inside the gasifying reactor, interconnecting one zone to another to understand the gasification processes' hydrodynamics better. A schematic representation of hydrodynamic cold flow developed for circulating fluidized bed riser (gasification of biomass) is shown in Fig. 4.12 [30]. This has been taken an example to prove a complete explanation in the current section. In principle, the cold flow experiment/system is isothermal and non-reactive. The major advantage of developing a cold flow system (experimented) is accumulating data to validate the actual hydrodynamic model, which is reactive and non-isothermal. The second is an estimation of the dynamic flow behavior to have a rough estimation for operating the reactor in a wide range of operating conditions and validating the future simulation of an industrial-scale circulating fluidized bed (CFB) riser. The figure shows that the

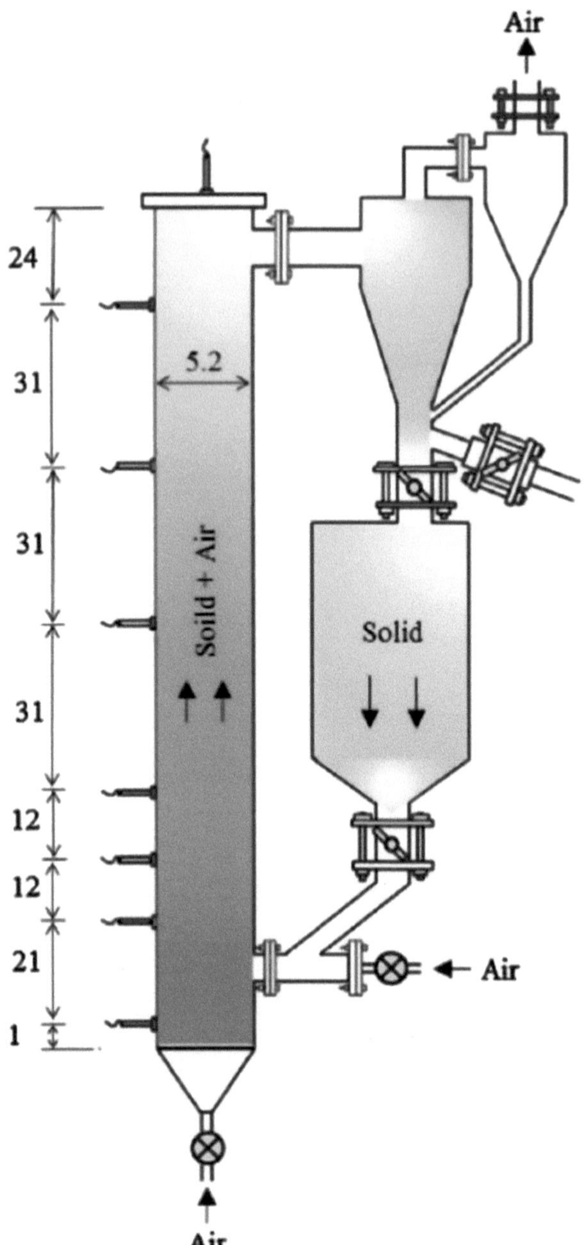

Fig. 4.12 Schematic representation of the cold flow circulating fluidized bed (dimensions are in cm) [30]

riser comprises glass fitted with pressure probes connected to pressure sensors/ transmitters. An auxiliary unit on the reactor's downside allowed airflow to the system. In addition, two rotameters were used to measure airflow into the circulating fluidized bed reactor. A set of predefined operating parameters (air flow rate, pressure transmitter, operating conditions, particle size, and density, fluidization velocity soli circulation rate, and fluidization medium) in a particular range in the dimension of circulating CFB reactor applied to cold flow model to a summary of results which will be implemented to hydrodynamic model CFB reactor. The finally appropriately obtained validated result will make the designed/developed model more robust and generic.

After experimenting with cold flow experiments, the designed hydrodynamic model (ANSYS workbench software) simulation geometry (shown in Fig. 4.13). While simulating the ANSYS hydrodynamic model of the CFB reactor, initially, the simulation domain was defined based on the atmospheric pressure. The solid and gas media are introduced to the model by specifying specified flow conditions. The result obtained from the simulation approach was analyzed through model sensitivity analysis. The model sensitivity analysis includes the transient behavior and time required to achieve a steady state value at different time steps, such as 0.1 s, 0.01 s, 0.001 s, etc. The above-defined hydrodynamic modeling approach is limited to

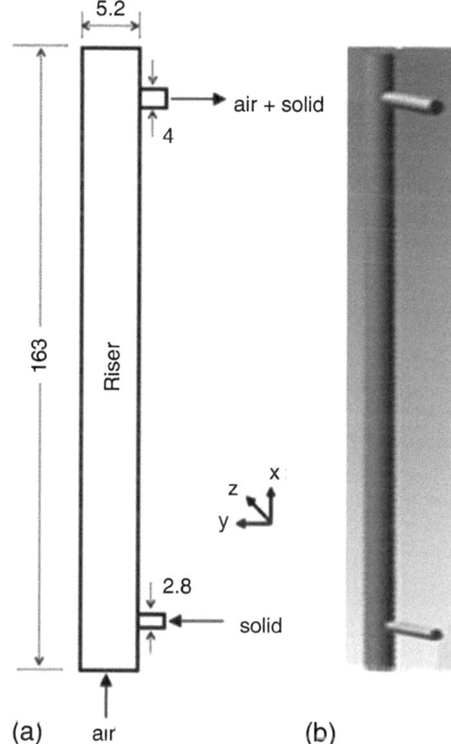

Fig. 4.13 Design and simulation of integrated gasification technology with other energy-driven systems (**a**) 3D geometry of the simulation domain generated by ANSYS Workbench software (dimension in centimeters) (**b**) [30]

modeling circulating fluidized bed reactors, but another gasification reactor can also be modeled to the desired condition. The intention here is to discuss the CFB reactor as an example of consideration for explaining hydrodynamic modeling.

4.4.3　Design and Simulation of Integrated Gasification Technology with Other Energy-Driven Systems

The integration of gasification technology with other energy systems and/or process equipment is primarily intended to derive effectiveness and to use the product of gasification technology in another system to produce energy and other useful chemicals. Generally, the integration of gasification technology defines the integrated gasification combined cycle (IGCC) process. This is because gasification technology-based gasification was used for power generation in the last decade. However, in recent years, the gasification plant can be integrated with other technologies to fulfill many objectives. Integrating gasification technology with other technologies could have particular challenges; however, the current progress in gasification technology and advanced ways of pursuing gasification have enabled efficient integration. In the current section, a rigorous discussion will be made on integrating gasification technology with technology/system, and the IGCC process will be less focused on as it has been mainly discussed in previous articles/books. Considering the aspect of sustainability of gasification technology and adaptation of renewable sources for energy generation, a discussion will be elaborated by taking an example of an integration of gasification technology with solid oxide fuel cell (SOFC) systems [31] and the integration of gasification combined cycle with chemical looping gasification (CLG) [32] for energy/power/heat/electricity generation and the integration of a combined cycle plant with a membrane reactor significantly impacts CO_2 capture and adsorption [33, 34].

A schematic outline of integrated gasification (biomass) with an SOFC system is displayed in Fig. 4.14 [31]. The combination of gasification technology (biomass, fossil fuel, etc.) with fuel cell systems (SOFC) is undergoing research and development. Integrating gasification systems with SOFC, mainly operated at higher temperatures, shows a promising way to develop a sustainable and highly efficient energy conversion system. The major parts of developing gasification-SOFC technology include the characterization of feedstock, thermochemical conversion inside the gasifier, and other operating parameters affecting the gasification process. In addition to the above, a significant factor that influences the overall process is impurities present in producer gas generated from the gasification process, and the impurities are tar, H_2S, HCl, and other alkali compounds. Therefore, the current focus is developing effective strategies for cleaning the producer gas before feeding it to the SOFC system. Overall, efforts are being made to develop SOFCs that can tolerate impurities in producer gas generated from gasification processes. Many theoretical

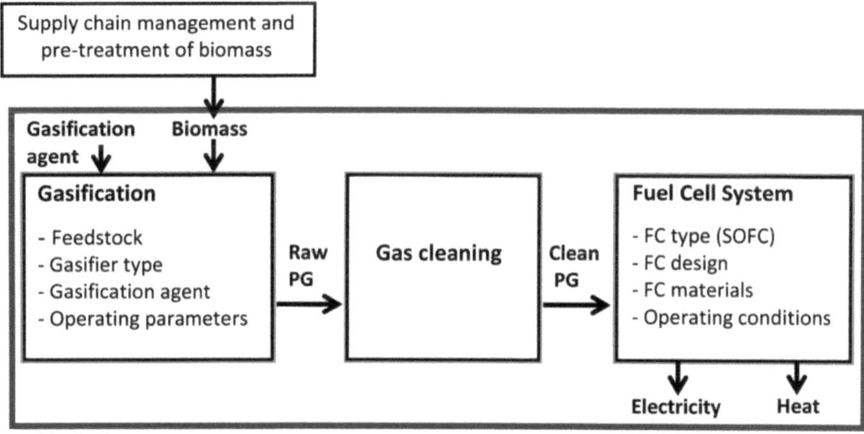

Fig. 4.14 Process outline of integrated gasification (biomass) with SOFC system [31]

and experimental studies are conducted in different research institutes and organizations.

In the case of the IGCC plant integrated with the chemical looping gasification (CLG) system (example of another integrated system approach), the production of power generation is enhanced compared to the individual operation of the IGCC plant. This is because in the presence of a CLG unit with IGCC, the syngas yield increased, and catalytic tar cracking occurred due to metal oxygen carrier material in CLG processes. That ultimately increases the rate of power generation. A schematic view with flowsheet of biomass IGCC-CLG system with hematite oxygen carrier shown in Fig. 4.15 [32]. According to experimental and theoretical investigations, the efficiency of the CLG-BIGCC plant is 33.51%, whereas the existing IGCC system has a power efficiency range of 30–32%. Therefore, the CLG-BIGCC system has a better priority for solid fuel to electricity generation, which leads to additional benefits for high-yield syngas production.

Moreover, the integration of the IGCC pant with a membrane reactor is primarily set for CO_2 capture and CO_2 removal from the producer gas generated by the gasifier. In the current approach, the IGCC plant is followed by a membrane reactor (e.g., catalytic palladium membrane reactor, CMR) to effectively remove CO_2 via CMR, where the producer gas enters the membrane reactor by generating two product streams. One product stream is CO_2-rich retentate, which goes further processing, i.e., CO_2 compression and liquefaction. The other product stream, gas without CO_2, passed to the steam cycle recovery unit to generate exhaust gas without CO_2. The detailed process is outlined in Fig. 4.16 [33]. Moreover, this kind of gasification integrated with membrane technology is emerging rapidly, and lab-scale experimentation shows that the integrated system's overperformance is more appealing than the increase in the performance of the membrane reactor.

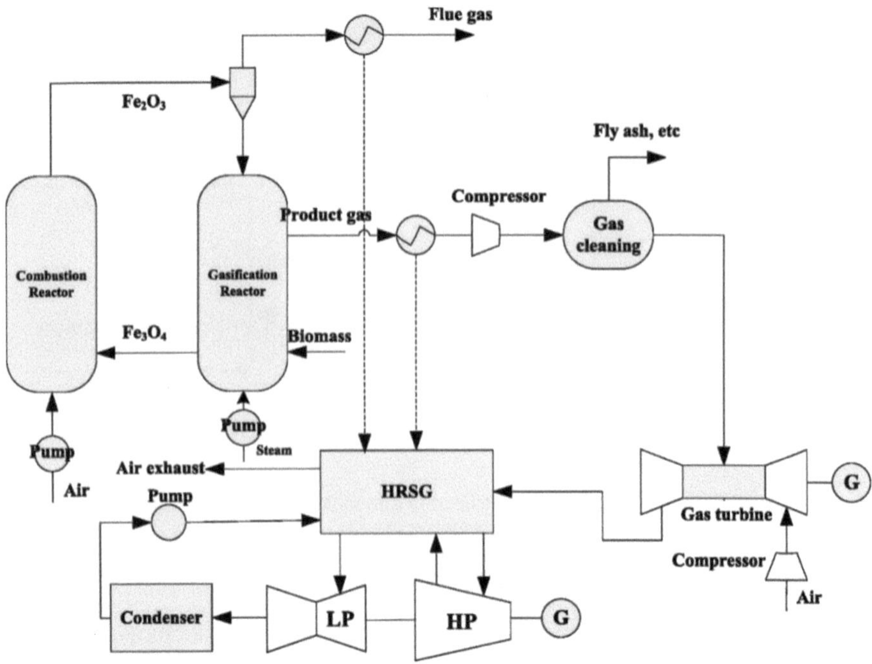

Fig. 4.15 Schematic diagram showing flowsheet of biomass IGCC-CLG system with hematite oxygen carrier [32]

In conclusion, the integration approach of gasification technology with commercialization to the industrial level leads to many considerations in the aspect of efficiency of the overall process, scaling up of technology from lab scale to industrial scale, and applicable to a wide range of feedstock for gasification. Also, while developing an integrated gasification technology, some approaches should be given the liability of process and sustainability in the current environment scenario.

4.5 Major Problems and Approaches

Although gasification technology has mainly been used across different countries, the change in current demand for energy and the availability of other renewable sources for feedstock have made gasification technology more challenging for efficient operation. A few decades ago, fossil fuel coal was the primary feedstock used for gasification, and biomass became the source for the gasification process. In recent years, biosolids, animal waste, agri waste, forestry residue, and other organic waste (food and garden organic waste) have been. The current trend of waste to energy and circular economy prospects has allowed a wide range of feedstock to solve many environmental problems, in the same way, that led to creating specific

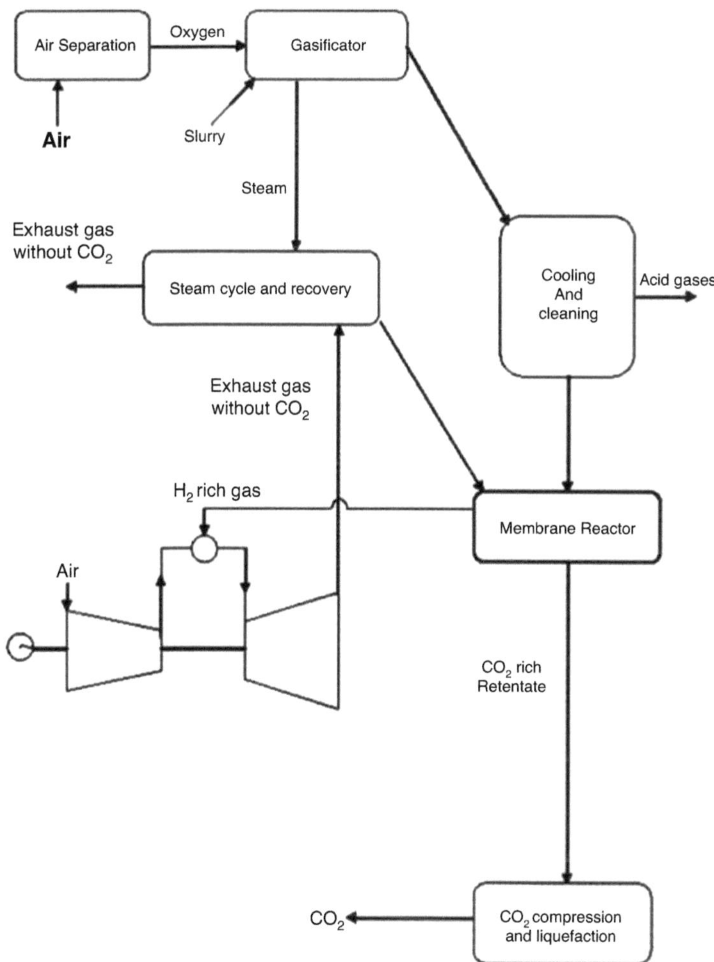

Fig. 4.16 Schematic diagram of IGCC plant integrated with the membrane reactor system [33]

challenges to optimize the process gasification process and other issues in dealing with the associated contamination in the feedstock. Furthermore, to generate a good quality of biochar for solid fuel (gasification process), using a single feedstock would not be sufficient to achieve all desired properties (higher fixed carbon and higher concentration catalytic AAEM species). Therefore, co-feeding is essential in some cases, as it adds two different types of feedstocks in different proportions based on synergistic effects. In the case of co-feeding, homogenizing the feed mixture into a single entity could lead to another challenging task, as different feedstock has different properties (density, moisture content, and other compositions). The new positive approach to the gasification process can lead to the formation of activated char precursor from char/biochar gasification in a particular environment (partially gamified char). It can be used as a catalyst and/or carbon sequestration for other thermochemical processes.

4.5.1 Use of Activated Char Precursor for Catalytic Reactions and Thermochemical Process

The gasification of char for energy production would not be the only objective that can lead to making the gasification technology more sustainable. Additionally, many approaches have been made to use the partially gasified char product as an activated char precursor for other applications. Those applications of activated char (partially gasified char in the particular gasifying environment) can act as catalysts for any solid-gas heterogeneous reaction. Also, it can act as an activated char precursor for other thermochemical reactions and char in another environment to maximize the desired product yield concentration.

The derived biochar (Mallee biomass, available in Western Australia) is prepared through pyrolysis and subsequent partial gasification. The gasification of pyrolyzed char was carried out for 5–10 min at a temperature of 750–850 °C to increase the oxygen(O)-containing functional group on the char's surface, further improving the catalytic activity [35, 36]. The studies of the biochar of Mallee biomass proved that partially gasified activated biochar is promising in reforming catalyst and catalyst support. The mechanistic view of reforming and decomposition reaction pathways, especially for catalytic steam reforming of bio-ethanol into hydrogen production, is shown in Fig. 4.17. The special characterizes of biochar as being functionalizing properties generated from inherent alkali and alkaline earth metallic species (AAEM) and the content of O-containing functional groups [37].

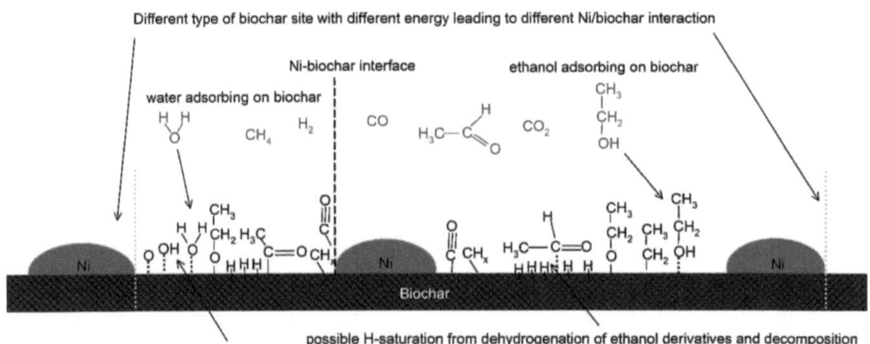

Fig. 4.17 Mechanistic view of ethanol steam reforming and decomposition reaction using Ni/biochar catalysts [37]

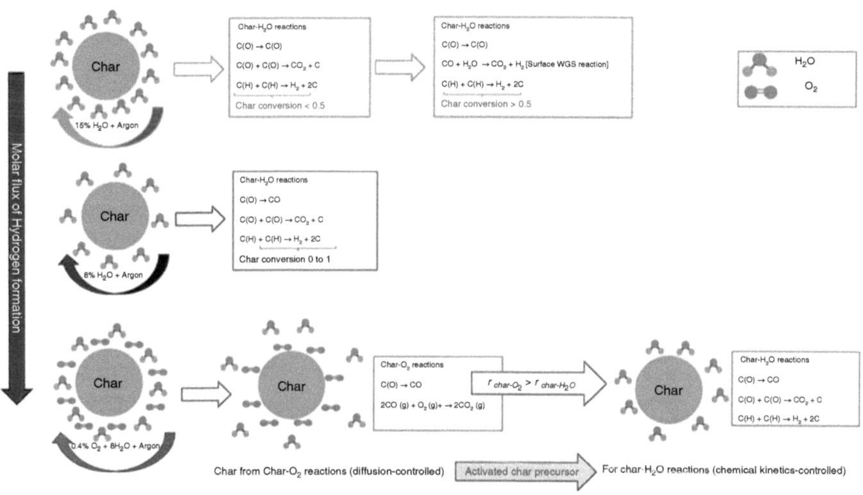

Fig. 4.18 Reaction mechanisms of Collie coal char (106–150 µm) gasification in the steam environment, i.e., 15%H₂O–Ar, 8%H₂O–Ar, and in the mixed gasifying environment, i.e., 0.4%O₂ + 8%H₂O–Ar [38]

Furthermore, partially gasified char in a gasifying environment can also be used as activated char solid fuel in other environments (Fig. 4.18). For example, partly gasified char produced from char-O_2 gasification reaction (only limited oxygen environment) might act as an active solid fuel for gasification char in a steam gasification environment. This may yield more hydrogen in the direct gasification of pyrolyzed char in a steam environment. It may be suggested that the partial gasification of pyrolyzed char in a limited oxygen environment must result in more active sites. This ultimately allows more H_2O molecules to be adsorbed and subsequent surface reactions in a steam gasification environment. However, this is a step-wise process and can be done in a batch manner. To make this mechanism an instantaneous and continuous process, adding limited oxygen with steam (mixed gasifying environment) would be beneficial in producing more hydrogen in product gas steam than gasification of pyrolyzed char in a steam environment alone. An experimental study on Collie sub-bituminous coal varying the gasifying environment and condition says that gasification of char (pyrolyzed char) in the temperature range 750 °C to 900 °C in a 0.4%O_2 + 8%H_2O-Ar (balanced argon) environment produces more hydrogen compared to gasification of char in 8%H_2O-Ar. This is because the O_2 starts to react with char faster than H_2O, as the char-O_2 is mechanistically diffusion-controlled, and char-H_2O is kinetically controlled in temperature ranges from 750 °C to 900 °C. Therefore, the fast reaction of char-O_2 and instantly formed char act as activated char fuel precursors for the subsequent char-H_2O reaction [38].

4.5.2 Design and Simulation of Co-gasification Technology

The rapid increase in demand for energy also concerns environmental issues, leading to the development of new technologies and other alternative energy sources [3, 19]. On the other hand, fossil fuel's limited availability brings biomass and other renewable sources to the limelight. However, gasification of single feedstock such as biomass is not always desirable. There are specific challenges associated with single feedstock gasification. Those include limited and periodic availability of certain feedstocks, low calorific value, undesirable physiochemical properties, cost, and storage space. Co-gasification technology was adopted to avoid problems/ issues associated with the gasification of a single feedstock and make it useful for all types of feedstocks. The operation of gasification technology is not as simple as mixing two or more feedstock and then doing gasification. Though the co-gasification can be performed in a similar reactor (fixed-bed, fluidized-bed, entrained-bed, etc.), many challenges are still there to feed a homogenous mixture to the reactor and other operating conditions as the composition, physical-chemical properties, and ash fusion temperature of different feedstock vary widely. A co-gasification technology will add fuel flexibility, but the co-gasification optimization method is vital for an efficient gasification process. There are some co-gasification optimization methods based on the type of feedstock, such as computational fluid dynamics (CFD), response surface methodology (RSM), propagation of error (POE), Box-Behnken Design (BBD), and central composite design (CCD) (as shown in Table 4.1). The optimization method of the co-gasification process can be achieved through the design and simulation of co-gasification technology.

The design of the co-pyrolysis setup plays a vital role in defining the overall performance. As shown in Fig. 4.19, the design of the co-gasification process contains different sections. The central functionalization zone consists of three zones: the upper combustion zone, the gasification zone where the outlet of product gas occurs, and the lower combustion zone. The gasifier is fed from the top steadily and conventionally (a mixture of different solid feedstocks). Specifically to the main gasifier, the feed sample flows down through the upper combustion zone into the

Table 4.1 Co-gasification studies using optimization techniques

Feedstocks	Co-gasification optimization methods	References
Coffee husks, forest, and vine pruning residues	Computational fluid dynamics (CFD), response surface methodology (RSM), and propagation of error (POE)	[39]
Sand-mixed dairy manure	Box-Behnken design (BBD) and central composite design (CCD)	[40]
Oil palm frond (OPF) and coconut shells (CS)	BBD	[41]
Catalyst-loading (CL) sawdust (SD) and catalyst-loading wood pellets	CCD	[42]
Can lignite and sorghum	BBD	[43]

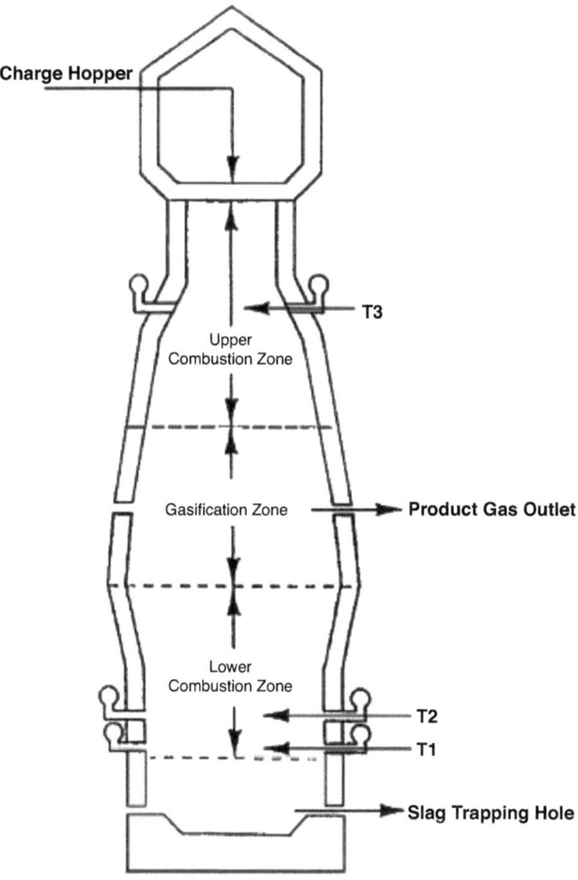

Fig. 4.19 Schematic view of co-gasifier [44]

gasification zone, where it comes in contact with oxygen, steam, and higher temperature gases, resulting in pyrolysis, gasification, thermal cracking of volatile component [44]. The significant reactions during co-gasification are similar to the gasification of single feedstock.

The simulation of co-gasification technology is nearly similar to the gasification technology/process simulation based on a single kind of feedstock [45, 46]. The simulation of a gasification model, hydrodynamic model, and thermodynamic process model has some limitations in terms of feedstock properties. Therefore, the co-gasification model, with varying feedstock properties, must be considered for all the fundamental aspects of gasification operations. For example, in the hydrodynamic modeling of co-gasification with significant variation in the density of primary and co-feedstock, the pelletization of feedstock would be the essential step for making the homogenous mixture [47]. Therefore, the design and simulation of a fluidized reactor would have been done in a manner that could have properly stimulated the fluidization of the pellet. Also, during continuous thermal operation, the feed pellets might disintegrate, changing the reactor's hydrodynamic behavior. An appropriate model could be designed and simulated to compensate for the effect of pellets and their disintegration. Despite the challenges associated with the co-gasification of pellets, additional focus must be given to integrating the co-gasification process with another pre-treatment method in the front-end co-gasification and other post-treatment processes. As illustrated in Fig. 4.20, the simulation of co-gasification technology is mainly subdivided into material processing, conversion processing, product, and use. Material processing is the first step in simulating a gasification process, which includes characterization (proximate, ultimate, and heating value) of feedstocks. The pre-treatment step must be carried out for both primary and co-feedstocks by mixing them before feeding to the gasifier. For example, as shown in the figure, the plastic and biomass waste were initially segregated for other required pre-treatment methods. Mixing feedstock to a homogenous mixture plays an essential role during a gasification reaction in a

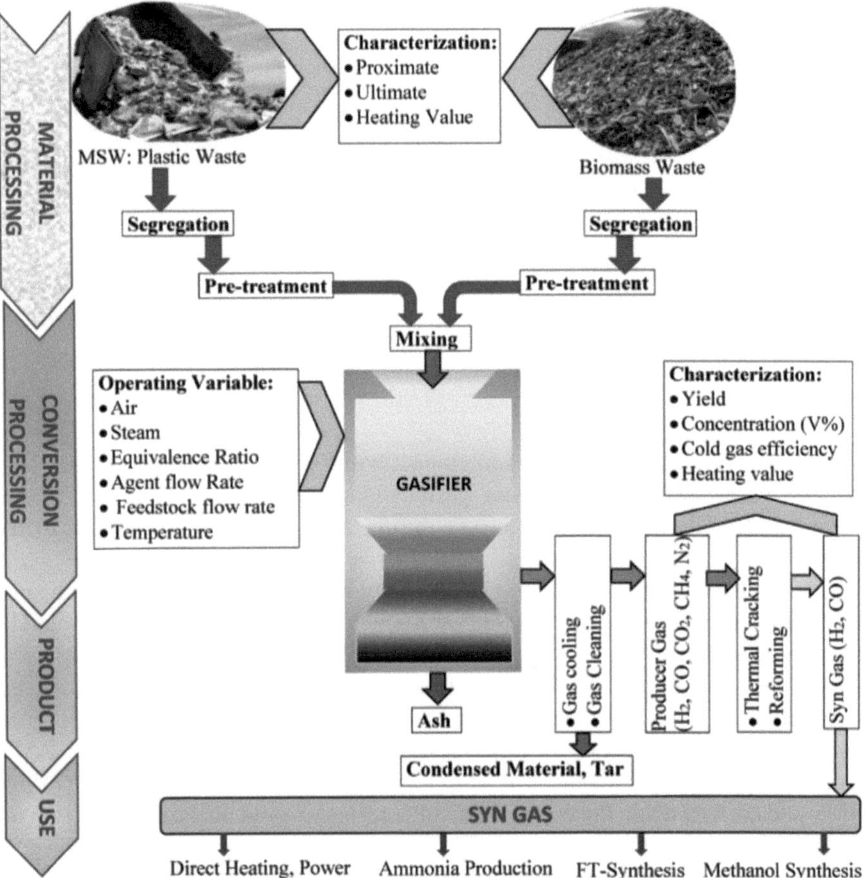

Fig. 4.20 Schematic diagram of co-gasification to product application [48]

hydrodynamic bed/operation. After the material process, there is conversion processing, another significant part of getting better and more efficient co-gasification operation. Many operating variables define the simulation of conversion processing. That includes the type of gasifying agent, equivalence ratio, gasifying agent flow rate, feedstock flow rate, and temperature of products, i.e., ash, condensed material, tar, and syngas. In addition, some conventional characterization needs to be focused on/monitored for an efficient operation; those are the product yield, the concentration of individual gases in the product gas mixture, cold gas efficiency, and heating value.

References

1. Ferreiro, A., et al. (2024). Influence of process parameters on biomass gasification: A review of experimental studies in entrained flow reactors and droptube furnaces. *Biomass and Bioenergy, 185*, 107217.
2. Rahim, M. R., et al. (2024). Gasification technology and its future: A review. *ASEAN Engineering Journal, 14*(1), 45–61.
3. Monteiro, E., Ramos, A., & Rouboa, A. (2024). Fundamental designs of gasification plants for combined heat and power. *Renewable and Sustainable Energy Reviews, 196*, 114305.
4. Senthilkumar, V., & Prabhu, C. (2024). Optimization of design and development of a biomass gasifier–a review. *Biofuels, 15*(8), 1–19.
5. Han, Z., et al. (2024). Major challenges and recent advances in characterizing biomass thermochemical reactions. *Resources Chemicals and Materials, 3*(2), 146–158.
6. Rasaq, W. A., et al. (2024). Navigating pyrolysis implementation—A tutorial review on consideration factors and thermochemical operating methods for biomass conversion. *Materials, 17*(3), 725.
7. Marcos, M., et al., Biochar for a sustainable EAF steel production (GREENEAF2). 2019.
8. Wang, L., et al. (2023). Synergistic effects and its influencing factors during co processing of coal and biomass in a wire-mesh reactor. *Journal of the Energy Institute, 108*, 101261.
9. Aniza, R., et al. (2024). Lignocellulosic biofuel properties and reactivity analyzed by thermogravimetric analysis (TGA) toward zero carbon scheme-a critical review. *Energy Conversion and Management: X, 22*, 100538.
10. Li, Y., et al. (2024). A review on thermogravimetric analysis-based analyses of the pyrolysis kinetics of oil shale and coal. *Energy Science & Engineering, 12*(1), 329–355.
11. Ogata, S., et al. (2004). Fluorination reaction of uranium dioxide by fluorine. *Journal of Nuclear Science and Technology, 41*, 135–141.
12. Rowell, R. & LeVan-Green, S., *6 Thermal Properties*.
13. Rowell, R., & Dietenberger, M. (2012). Thermal properties, combustion, and fire Retardancy of wood. In R. Rowell (Ed.), *Handbook of Wood Chemistry and Wood Composites* (pp. 127–149). CRC Books.
14. Asadullah, M., et al. (2009). Importance of biomass particle size in structural evolution and reactivity of char in steam gasification. *Industrial & Engineering Chemistry Research, 48*(22), 9858–9863.
15. Akhtar, M. A. (2020). *Biochar gasification mechanism*. Curtin University.
16. Liu, Y., et al. (2020). Difference in tar reforming activities between biochar catalysts activated in H_2O and CO_2. *Fuel, 271*, 117636.
17. Singh, S., et al. (2022). An integrated two-step process of reforming and adsorption using biochar for enhanced tar removal in syngas cleaning. *Fuel, 307*, 121935.

18. Breault, R. W. (2010). Gasification processes old and new: A basic review of the major technologies. *Energies, 3*(2), 216–240.
19. Yek, P. N. Y., et al. (2024). Co-processing plastics waste and biomass by pyrolysis–gasification: A review. *Environmental Chemistry Letters, 22*(1), 171–188.
20. Li, X., et al. (2024). Pyrolysis-free covalent organic polymers directly for oxygen Electrocatalysis. *Accounts of Chemical Research,* 12567–12572.
21. Zhang, L., et al. (2015). Structural transformation of nascent char during the fast pyrolysis of mallee wood and low-rank coals. *Fuel Processing Technology, 138,* 390–396.
22. Zhang, L., et al. (2015). Changes in nascent char structure during the gasification of low-rank coals in CO2. *Fuel, 158,* 711–718.
23. Javaid, S. F., et al. (2024). Production of biochar by slow and solar-biomass pyrolysis: Focus on the output configuration assessment, adaptability, and barriers to market penetration. *Arabian Journal for Science and Engineering,* 1–20.
24. Dafalla, M., et al. (2024). Prospective of biochar material production and process optimization using co-pyrolysis approach-a mini-review. In *Journal of Physics: Conference Series.* IOP Publishing.
25. Patra, B. R., et al. (2021). Slow pyrolysis of agro-food wastes and physicochemical characterization of biofuel products. *Chemosphere, 285,* 131431.
26. Phounglamcheik, A., et al. (2020). Effects of pyrolysis conditions and feedstocks on the properties and gasification reactivity of charcoal from woodchips. *Energy & Fuels, 34*(7), 8353–8365.
27. Francois, J., et al. (2013). Detailed process modeling of a wood gasification combined heat and power plant. *Biomass and Bioenergy, 51,* 68–82.
28. Materazzi, M., et al. (2013). Thermodynamic modelling and evaluation of a two-stage thermal process for waste gasification. *Fuel, 108,* 356–369.
29. He, C., Feng, X., & Chu, K. H. (2013). Process modeling and thermodynamic analysis of Lurgi fixed-bed coal gasifier in an SNG plant. *Applied Energy, 111,* 742–757.
30. Hassan, M. I., & Makkawi, Y. T. (2018). A hydrodynamic model for biomass gasification in a circulating fluidized bed riser. *Chemical Engineering and Processing – Process Intensification, 129,* 148–161.
31. Ud Din, Z., & Zainal, Z. A. (2016). Biomass integrated gasification–SOFC systems: Technology overview. *Renewable and Sustainable Energy Reviews, 53,* 1356–1376.
32. Ge, H., et al. (2019). System simulation and experimental verification: Biomass-based integrated gasification combined cycle (BIGCC) coupling with chemical looping gasification (CLG) for power generation. *Fuel, 241,* 118–128.
33. Amelio, M., et al. (2007). Integrated gasification gas combined cycle plant with membrane reactors: Technological and economical analysis. *Energy Conversion and Management, 48*(10), 2680–2693.
34. Gazzani, M., Macchi, E., & Manzolini, G. (2013). CO_2 capture in integrated gasification combined cycle with SEWGS – Part A: Thermodynamic performances. *Fuel, 105,* 206–219.
35. Min, Z., et al. (2011). Catalytic reforming of tar during gasification. Part II. Char as a catalyst or as a catalyst support for tar reforming. *Fuel, 90*(7), 2545–2552.
36. Min, Z., et al. (2013). Catalytic reforming of tar during gasification. Part IV. Changes in the structure of char in the char-supported iron catalyst during reforming. *Fuel, 106,* 858–863.
37. Afolabi, A. T. F., et al. (2021). Kinetic features of ethanol steam reforming and decomposition using a biochar-supported Ni catalyst. *Fuel Processing Technology, 212,* 106622.
38. Jena, M. K., Kumar, V., & Vuthaluru, H. (2022). Investigation into kinetic compensation effects for the production of hydrogen-rich gas during gasification of sub-bituminous coal char in varying gas environments. *International Journal of Hydrogen Energy, 47*(89), 37760–37773.
39. Silva, V., & Rouboa, A. (2015). Combining a 2-D multiphase CFD model with a response surface methodology to optimize the gasification of Portuguese biomasses. *Energy Conversion and Management, 99,* 28–40.

40. Nam, H., et al. (2016). Enriched-air fluidized bed gasification using bench and pilot scale reactors of dairy manure with sand bedding based on response surface methods. *Energy, 95*, 187–199.
41. Inayat, M., et al. (2020). Application of response surface methodology in catalytic co-gasification of palm wastes for bioenergy conversion using mineral catalysts. *Biomass and Bioenergy, 132*, 105418.
42. Mansur, F. Z., et al. (2020). Co-gasification between coal/sawdust and coal/wood pellet: A parametric study using response surface methodology. *International Journal of Hydrogen Energy, 45*(32), 15963–15976.
43. Secer, A., & Hasanoğlu, A. (2020). Evaluation of the effects of process parameters on co–gasification of Çan lignite and sorghum biomass with response surface methodology: An optimization study for high yield hydrogen production. *Fuel, 259*, 116230.
44. Yuehong, Z., Hao, W., & Zhihong, X. (2006). Conceptual design and simulation study of a co-gasification technology. *Energy Conversion and Management, 47*(11), 1416–1428.
45. Fajimi, L. I., Oboirien, B. O., & Adams, T. A., II. (2024). Waste Tyre gasification processes: A bibliometric analysis and comprehensive review. *Fuel, 368*, 131684.
46. Shen, Y. (2024). Biomass pretreatment for steam gasification toward H2-rich syngas production–an overview. *International Journal of Hydrogen Energy, 66*, 90–102.
47. Wan, Q., et al. (2024). The application of process simulations in lignocellulosic biorefinery: A review. *Biofuels, Bioproducts and Biorefining, 18*(2).
48. Raj, R., Singh, D. K., & Tirkey, J. V. (2023). Co-gasification of plastic waste blended with coal and biomass: A comprehensive review. *Environmental Technology Reviews, 12*(1), 614–642.

Chapter 5
Sustainability and Prospects for Gasification Technology

Paving the Way for Sustainable and Cleaner Energy Solutions for the Future Through Transformation

5.1 Introduction

In the last few decades, gasification technology has been proven to be efficient in fulfilling energy demand using a wide range of feedstocks. Gasification technology and processes produce heat and power from products such as biofuel/fuel and syngas. Despite the many advantages of the gasification process, specific challenges create barriers to domestic and commercial use of this technology [1]. Those challenges and barriers globally include feedstock/solid fuel supply chain management, pretreatment of feedstocks, and additional problems in gas conditioning and conversion technology. However, the role of scientific communities in addressing these challenges and improving gasification technology is crucial. Furthermore, based on the study on a regional basis (different countries and states), the government policies and utilization of gasifier products such as flue gas for heat and power generation have been the biggest and continued factors associated with commercialization technology [2, 3]. Many factors mainly determine the sustainability of a gasification technology. To ensure the gasification process has a sustainable future, generic barriers need to be eliminated. Those barriers are classified into institutional, informational, financial, policy, and market. Hence, this chapter explains the modernization of gasification technology, the state of the art in the industrial application of gasification technology, and the fate and role of gasification technology for the next generation to summarize the sustainability of gasification processes. Also, the role and responsibilities of scientific communities for gasification technology protection will finally be provided to assess a better future for this technology in the upcoming years.

M. K. Jena, H. B. Vuthaluru, *Gasification Technology*,
https://doi.org/10.1007/978-3-031-71044-5_5

5.2 Modernization of Gasification Technology

The modernization of gasification technology is the advancement of gasification technology and processes to make the process more efficient and sustainable than other thermal technologies [4]. The gasification technology is primarily focused on the thermochemical conversion of solid fuel. The approach can include thermochemical and non-thermochemical conversion to make the gasification process more efficient [5]. The thermochemical conversion approach includes optimizing process parameters, cleaning syngas as a product, recovering more heat and power, and using a wide range of feedstock. The non-thermal process includes exploring pre-treatment processes such as steady and continuous feeding of different samples, the effect of co-feeding during co-gasification processes, and strategies for hot gas cleaning and efficient heat recovery [6]. Therefore, the scientific and industrial communities have developed many effects and approaches to achieve those objectives. A brief discussion will be provided in the following sections.

5.2.1 Advances in Waste Management Towards Gasification Technology

The current world of continuous increase in energy demand and depletion of fossil fuels has forced us to find sources of alternatives. In the same period, the continuous increase in the Earth's population has strived to make a clean environment by effectively utilizing waste. Effective utilization of solid waste and fulfilling the demand for energy provides the platform for advancing the waste management scheme toward the gasification technology [1, 7]. In a scenario of the current availability of many gasifiers (e.g., entrained flow gasifier, downdraft gasifier, updraft gasifier, bubbling fluidized bed gasifier, circulating fluidized bed gasifier, rotary kiln gasifier, and plasma gasifier) as well as the validity of different solid carbonaceous fuels. Even though the recent approach to solid waste management is moving towards the trend of a hybrid approach, which combines gasification technology with other thermal technology. This approach enables an increase in the capacity of overall technology and equipment for better product and energy generation. The prime objective of coupling gasification technology with other technology is to increase the efficiency of the overall process and release less harmful gases to the environment [6, 8]. Some approaches have been made to developing gasification technology integrating with other technology for efficient product recovery and power generation, as shown in Table 5.1. In the current situation, the continuous rise in population growth is directly related to the improper use of waste and other renewable sources. Gasification is a primary and economical approach to converting solid waste and other renewable sources into profitable, nonhazardous products and gases. However, a techno-economic assessment is necessary to understand the influence of gasification technology and its hybrid technology in carrying with low environmental

Table 5.1 Gasification and its hybrid technologies

Primary technology	Secondary technology (integrated with primary technology)	Major advantages	Reference
Fluidized bed gasification	Solar thermal technology	Vegetable residue can be recovered more economically and efficiently from raw materials like almond shells, olive trees, and oak pruning	[9]
Endothermic pyrolytic gasification	Nuclear reactor	Less cost for production of diesel and power from municipal and agricultural waste	[10]
Fixed bed downdraft gasifier	Anaerobic digestion system	Increase the yield of energy production from solid waste	[11]
Downdraft gasifier	Micro gas turbine and solid oxide fuel cell (SOFC)	Higher efficiency in the production of thermal and electric power from biomass material	[12]

impact conditions. The techno-economic assessment suggests that among all the defined hybrid technologies (Table 5.1), around 71% process efficiency is achievable in the case of integrated aerobic digestion with a fixed bed gasifier. However, solar thermal technology and nuclear technology, in integration with the gasification process, have a higher yield of oil and energy generation.

In case of gasification, an integrated SOFC system, around 1 MW of power is required for efficient conversion [13, 14]. Despite all that, a gasification technology may not be attractive in terms of capacity (space/area), and operation costs will be higher than other thermo-chemical conversion technologies. However, concerning the environmental assessment, the gasification technology could provide a better opportunity for the thermochemical conversion of solid fuel. Therefore, the representative life cycle assessment is also essential in identifying the potential environmental impact of other waste management technologies, including gasification technology [6]. According to the assessment, apart from solving technical challenges in expanding and commercializing waste management technology, the primary intention should be to focus on the environmental influence of raw materials and the disposal of thermochemically converted residue and ashes [15]. Hence, to solve those raised problems, integrating the gasification process with other technologies in front and back of gasification (Figs. 5.1, 5.2 and 5.3) can partly solve the environmental waste management problems. In an approach, still, lab scale and pilot scale trials and research are expected to solve the challenges associated with a wide of waste (argi waste, municipal waste, animal waste, pulp and paper waste, and other forestry residues), availability of feedstock, government policies and hazardous effects during operation.

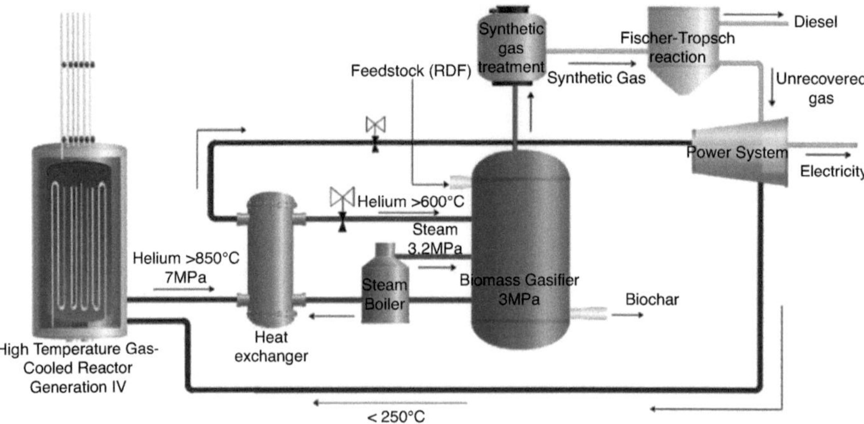

Fig. 5.1 Schematic flow of biomass-nuclear hybrid system [10]

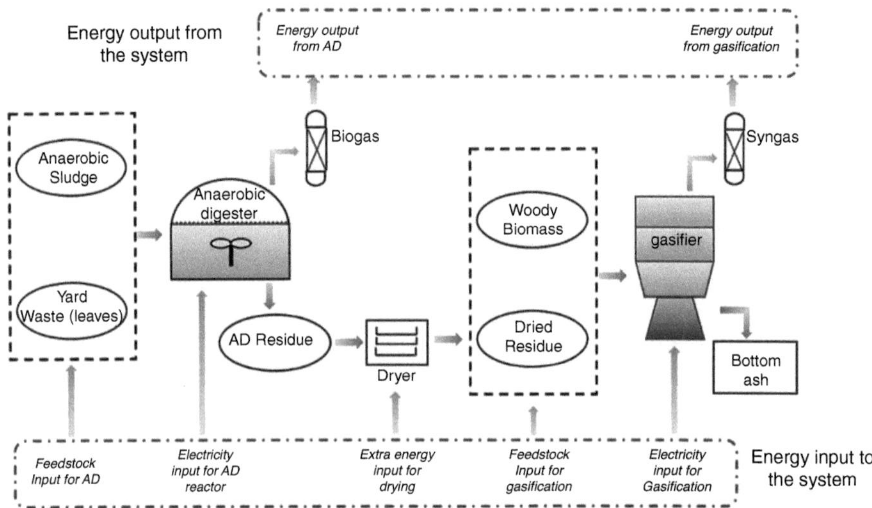

Fig. 5.2 Process flowsheet of the two-stage hybrid system [16]

5.2.2 Current Practice and Prospects for Value-Added Chemicals

The thermochemical conversion of technologies provides a sustainable path for integrating solid fossil fuel and other solid renewable sources to produce valuable oil, gaseous, and solid products. The thermochemical conversion of any solid fuel usually occurs through pyrolysis, gasification, and torrefaction, which are interrelated. Therefore, it is difficult to state that the conversion of solid fuel to chemical through gasification rather than thermochemical conversion [18]. The major

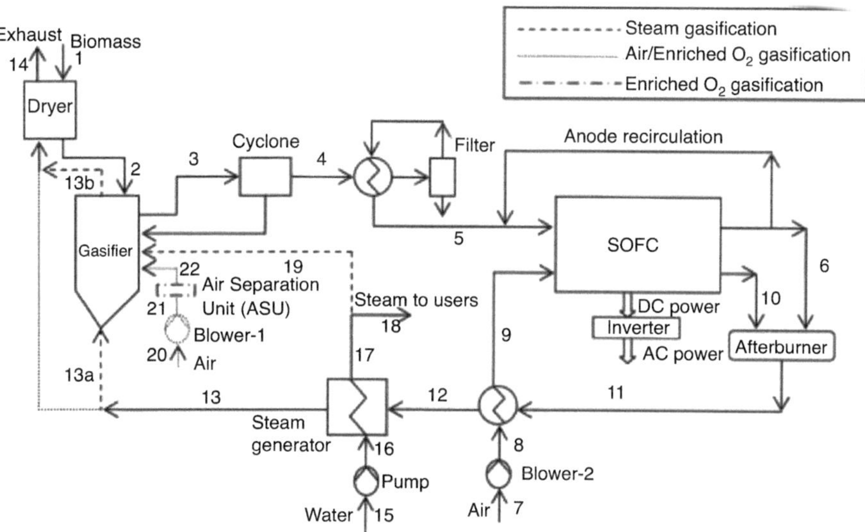

Fig. 5.3 Schematic of the integrated biomass gasification and SOFC system [17]

advantage of thermochemical conversion of solid fuel over another thermal treatment method (incineration) is to produce value-added products in an increased energy-efficient manner with an improved pollution control strategy [19]. In an overall thermochemical conversion process, the derived bio-oil/tar and product gas can be used directly to convert into value-added chemicals. Meanwhile, the solid product char/bio-char, usually obtained after partial gasification, can participate in another subprocess as catalysts/adsorbents/sequestrated carbon material for another chemical process for producing valuable chemical products.

In specifying gasification as a key conversion technology in converting solid fuel into chemicals, the major factors achieving this objective include the feedstock type, operating parameters, type of reactor, and product yield. An additional challenge associated with solid fuel gasification is the formation of tar substance, which could bring other issues such as fouling and corrosion. Therefore, in the recent approach technique, tar removal strategies are applied by some physical methods, such as filtering and scrubbing, as well as chemical processes like thermal and catalytic cracking. In the above-mentioned strategies for removing the tar, additional approaches are given to the gasification process to generate less tar from the thermal operation. One of the recent approaches is steam gasification of solid fuel (e.g., municipal solid waste, MSW) with the occurrence of an exothermic reaction, which shows reduced tar formation with an increased yield of H$_2$ in the product gas [20]. Also, adding an external catalyst (CaO) to the process and CO$_2$ sorbent in downstream operation (Fig. 5.4) could lead to substantial tar decomposition at a higher temperature [21].

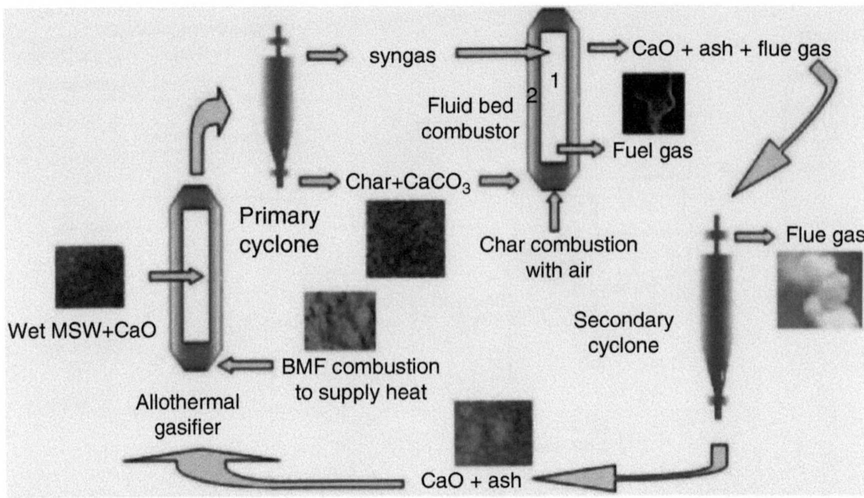

Fig. 5.4 Gasification of wet MSW using CaO as a catalyst and CO_2 sorbent in the downstream operation [21]

5.2.3 Exploration and Improvement of Gasification Technology Towards Sustainability

The exploration and improvement of gasification technology would not be an approachable path of study as long as it was not studied in conjunction with individual applications along with another process. In some cases, strategies have been in place to form an advanced gasification reactor and integrate gasification technology as a sub-process with other processes [22]. Also, forming a generic technology for prioritizing the gasification process with a wide range of feedstocks would be a way to achieve the sustainability of gasification technology [23]. In the current growing world, the co-gasification process has been largely implemented at lab- and pilot-scale levels. This resulted in the efficient production of high-value syngas. Also, efforts were made to diversify the feeding of gasifying agents in various temperatures and compositions. A list of approaches that have been adopted in the current scientific community to make gasification technology more sustainable in the last few decades is detailed below [24, 25]:

- Gasification of biomass for sustainable energy production and with chemical looping technology for CO_2 capture.
- Gasification of municipal solid waste blended with other renewable sources to develop hybrid technologies and innovative sources.
- Gasification technology is integrated with SOFC systems for efficient and renewable power generation.
- Enhancement of process in gasification technology for enriching the production of hydrogen from coal: an economical way to sustainable energy.

- Advancement in developing plasma gasification, supercritical water gasification, and microwave-assisted gasification for sustainable gasification technologies.

In developing gasification to forecast a sustainable future, emphasis was also given to the economic and social impact of processes. In many cases, individual feedstock approaches are being made, such as gasifying municipal solid waste, biosolids, agri waste, forestry residue, and animal waste for commercialization. However, gasification technologies based on single feedstock feeding would have faced many challenges [26]. These could be related to social issues, improper availability of feedstocks, local government policy, and other issues being linked to cost management and contamination associated with the feedstocks.

5.3   State of the Art in Industrial Application

Understanding gasification technology is very crucial in specifying many industrial applications. Gasification technology has a wide range of applications. At the beginning of power generation from coal gasification, in recent times, there has been a wide range of applications of the gasification process [27]. From the prospects of utilizing a wide range of feedstocks, the gasification process can be used in many industries. Also, as an integration part of another process, the gasification process can be diversified into many fields. Some examples where the gasification process has been improved include: co-gasification of different feedstocks, biomass and solar hybrid gasification followed by solar-pyrolysis, supercritical water gasification, circulating fluidized bed gasification, and the use of plasma reactor for gasification [28]. Based on the current approach, this section explains the status of supercritical water gasification and microwave-assisted gasification. Also, the industrial approach of a plasma reactor for gasification and the scope of gasification technology as an emerging path for hydrogen production will be discussed further. Finally, a brief elaboration on the recent trends in demand for gasification technology for industrialization will be covered.

5.3.1   Current Status in Supercritical Water Gasification and Microwave-Assisted Gasification

As to the current work on hydrothermal gasification, supercritical gasification provides an alternative pathway for feedstock gasification with a water content of up to 80%. In a supercritical gasification, the feedstock decomposes within a minute or second. Supercritical water gasification utilizes water as a solvent in a supercritical condition. The operating conditions are about greater than the supercritical condition of water (approximated around 374 °C and 22.1 MPa) [29]. Under supercritical conditions, water behaves in gas and solid-like properties, which leads to unique properties that benefit supercritical water gasification, such as low mass transfer limitations, high reaction rates, and higher quality of product gas. Supercritical water gasification has been implemented with a broad range of feedstocks, such as agriculture biomass, forestry residue, black liquor, algae, food waste, etc. [30–34].

Although many feedstocks have been tried and implemented for supercritical water gasification, the optimization of process parameters must be carried out before commercializing the industrial scale from the pilot scale. In addition, it would be very difficult to optimize the process considering the physicochemical properties of each feedstock, but some efforts are given to particular kinds of feedstocks. For example, the lab scale experimental study done for supercritical water gasification of canola straw at an optimized temperature of 350 °C to 500 °C for a reaction time of 10–28 min with a feedstock concentration of 10–25% displayed the formation of high-value H_2 gas mixture [29]. Also, the hydrochar obtained at high gasification temperatures revealed a stable aromatic carbon structure. Figure 5.5 presents the continuous view of super-critical water gasification with a post-critical reagent injection and in situ monitoring. It should be noted that when designing a common reactor for supercritical gasification operation, the primary focus should be on the front end or pre-treatment process before the main gasification operation. Overall, as discussed above, it was concluded that supercritical water (SCW) conversion technology could be one of the promising in-situ conversion techniques in benefiting a high conversion ratio with a yield of more desired hydrocarbons in an excelled/optimized heat and mass transfer performance.

In comparison to biomass, the supercritical water gasification of coal is a bit complicated as few more intermediate steps are involved. Some of the previous

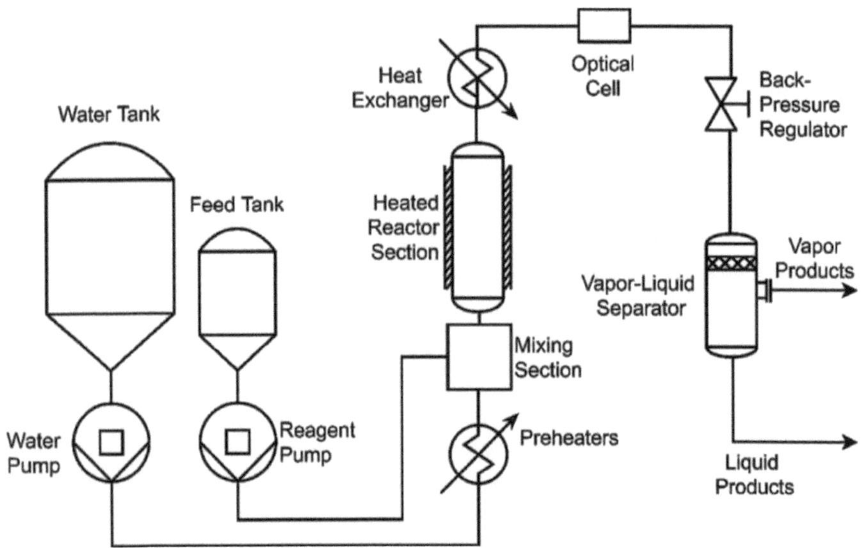

Fig. 5.5 Schematic view of supercritical water gasification [35]

studies reported (Fig. 5.6) that supercritical water gasification occurs through two competitive reaction steps. Among them is the ionic step, where the dominant reaction is controlled under high temperatures and low pressure. Other one is identified as a radical reaction, which is operated under high temperature and low pressure [36]. As can be depicted from Fig. 5.6, during the SCW, coal gasification containing various elements such as N, S, As, and Hg is deposited as organic salt because of the characteristics of supercritical water. The coal molecule undergoes hydrolysis, pyrolysis, extraction, and liquefaction reactions, resulting in liquid intermediates, fixed carbon, and product gases (H_2, CO, CH_4, CO_2, etc.). The as-formed fixed carbon gets further reacted, producing H_2 and CO through a reforming reaction with some unreacted char.

In a supercritical condition, the cellulose, hemicellulose, and lignin contained in biomass (wheat straw) undergo hydrolysis in the formation of glucose, xylose, and phenolics (approximated around T > 374C and P > 22. Mpa) [38]. The gasification pathway of biomass is shown schematically in Fig. 5.7. A further process of major reactions such as cellulose hydrolysis, glucose reforming reaction, lignin hydrolysis, monomer decomposition, and lignin steam reforming reaction leads to the formation of acids, alcohols, aromatics, aldehydes, and phenols. In addition to the mentioned reactions, some intermediate reactions occurred in the completion of the gasification process with the formation of CO, CO_2, H_2, and CH_4 through water gas shift and methanation reactions. Overall, the supercritical gasification mechanisms differ in the composition of the feedstock.

Based on the heating rate using a microwave, microwave-assisted gasification can be a promising technology for the conversion of biomass, coal, biosolids, and other carbonaceous materials. A process flow and simple experimental setup for microwave-assisted gasification are shown in Fig. 5.8a–b. Microwave-assisted gasification technology provides a broad platform for controlling microwave power,

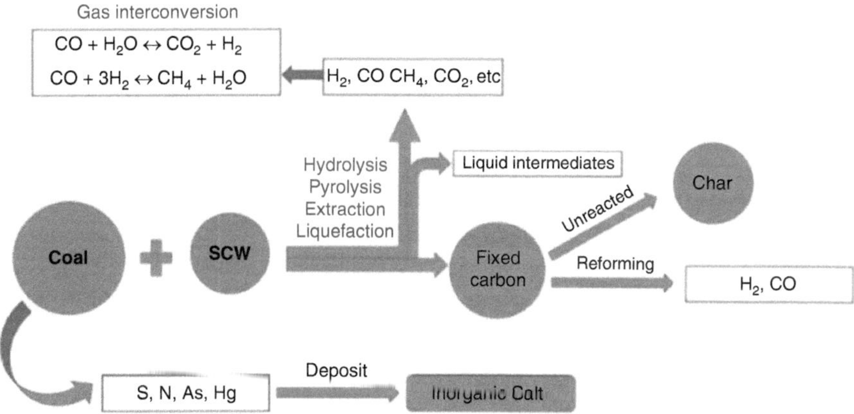

Fig. 5.6 Reaction mechanisms and pathways of coal conversion in supercritical water conditions [37]

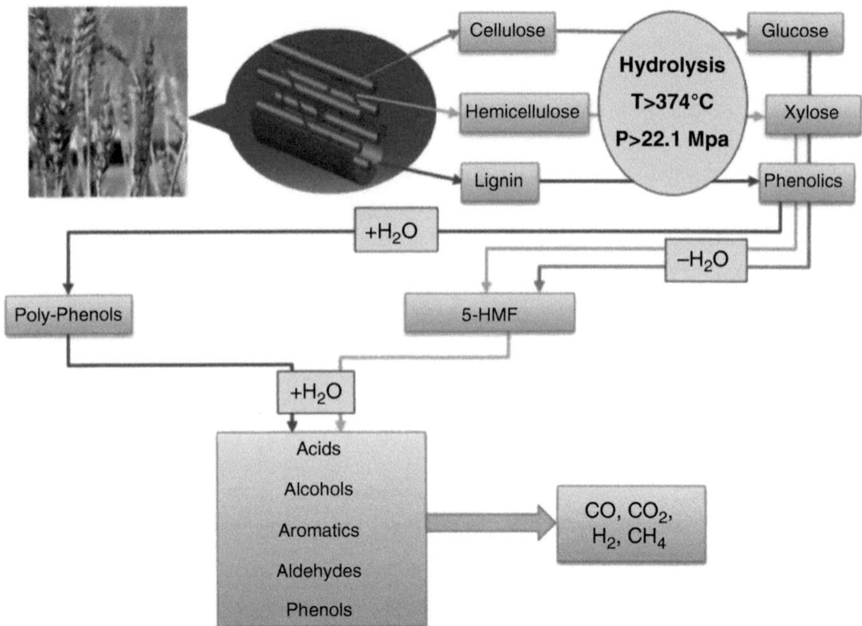

Fig. 5.7 Schematic pathways for supercritical gasification of biomass (wheat straw) [38]

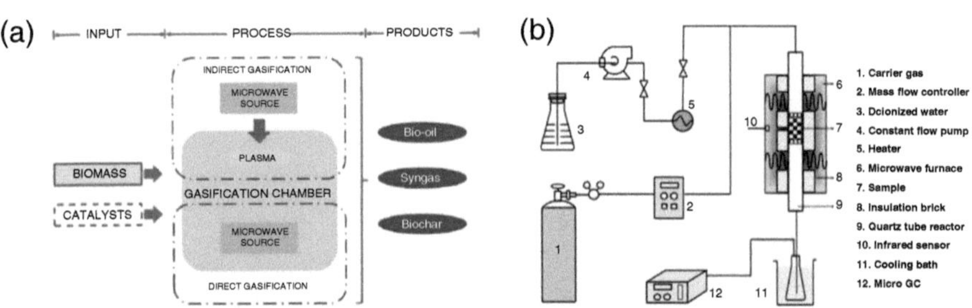

Fig. 5.8 A schematic microwave-assisted gasification system of process **(a)** and experimental setup **(b)** [41].

time, and conversion level to increase overall efficiency and optimize process conditions. So, in the scientific community, a limited number of researchers have reported microwave-assisted gasification technology. However, it was noticed that this kind of gasification technology could provide a sustainable production of biohydrogen and bio syngas via gasification [39]. Upon subjecting the heating source, gasifying agent, and nature of the gasification reaction, there are two major types of gasification: direct gasification and indirect gasification. Direct gasification occurs in a restricted amount of oxidant with proper utilization of the heat generated from exothermic oxidation reaction for thermal degradation of biomass [40]. On the other hand, the indirect gasification process does not use any oxidant as a requirement for an external energy source for heat transfer and electric mean. In comparing both types of gasification, microwave-assisted gasification is an indirect gasification that has the big advantage of replacing the conventional heating source. It also benefits from uniform volumetric heating and rapid thermal reactions.

Microwave-assisted gasification is one of the promising conversion techniques for achieving a sustainable economy. This technology shows massive potential for selective and effective biohydrogen and bio syngas production from biomass and other biological sources. The potential future of microwave-assisted gasification needs to be enlarged in the direction of large-scale industrialization by understating the general fundamental mechanism of microwave-assisted heating applied to gasification.

5.3.2 Industrial Aspects of the Plasma Reactor for Gasification

Plasma gasification is a thermal process where external power is used to heat the reactor and to keep a high temperature during operation. During plasma gasification operation, the material was decomposed in a limited oxidant environment, and higher conversion efficiency was achieved because of the higher operating temperature. The major products obtained are syngas slag and ash [42]. Plasma at higher temperatures breaks all the material into its elemental form. Therefore, it can be suggested that the plasma gasification technique can be operated in a wide range of feedstocks, excluding the radioactive substances, as it could lead to hazardous problems. Specifically, higher operation temperatures cause the decomposition of toxic compounds into harmless and/or less toxic compounds, and this provides major advantages compared to other conventional methods for gasification [25]. The primary factors segregating plasma gasification from others are plasma discharge technique and reactor characteristics.

Considering the enlargement of feedstock diversification that can be utilized for plasma gasification, this gasification technique shows a larger potential in industrial applications, as shown in Fig. 5.9 [43]. For example, municipal solid wastes, hazardous waste from oil and gas industries, and other sources, including agri, forestry, pulp paper, etc. It can be stated that the plasma gasification technique will not only solve the problem of producing heat and energy effectively. However, it also helps to create a possible path of destruction of pathogens and other contaminants, such as PFAS, present in feedstock during gasification [44]. Although this understanding of destructing the contaminants inside the plasma reactor may need more investigation, solving that kind of problem would enable us to achieve a solution that causes a social and environmental impact. Overall, if it is possible to destroy the contaminants during gasification, the additional cost of destructing the contaminants through thermal and non-thermal processes could be stepped out.

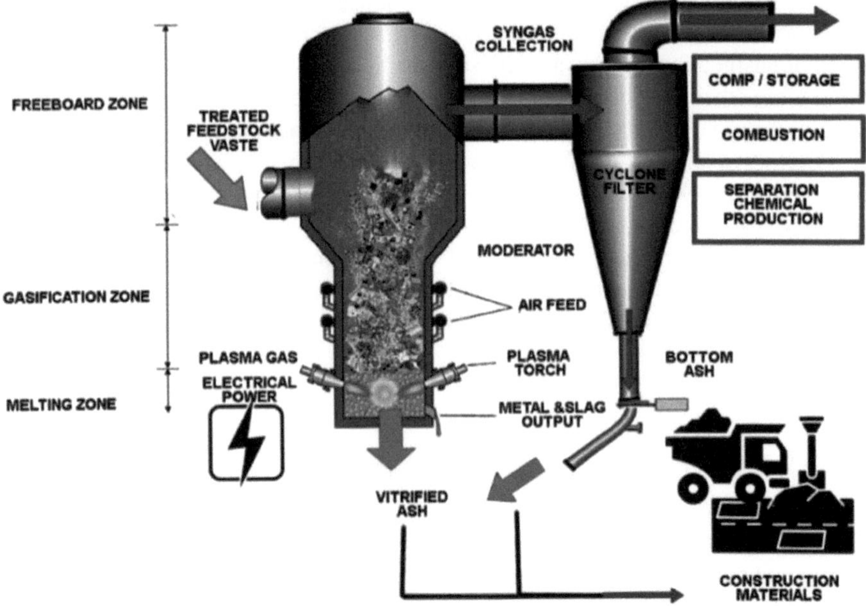

Fig. 5.9 Plasma gasification setup with the utilization of downstream for further processing [43]

5.3.3 Scope of Gasification Technology as a Direct Path for Hydrogen Production

The gasification process is not only a path for generating high-value syngas and energy, but it could also be an economical option for generating high-value hydrogen. To make gasification technology a sustainable future, the primary objective of producing hydrogen could be a promising one. However, hydrogen production from a gasification process primarily depends on feedstock type, gasifying agent, and operation conditions. Also, in certain cases, the type of gasification plays a major role in producing enriched H_2 in the product gas. Many approaches have been made to produce hydrogen through gasification, and some of the new approaches are listed below.

- Intensified two-stage gasification system for production of hydrogen from biomass [45].
- Production of hydrogen from municipal solid waste and food waste through gasification [46, 47].
- Production of hydrogen through supercritical water gasification of biomass [48].
- Hydrogen production from agro residue through gasification and adsorptive separation [49].

Since the gasification leads to the production of syngas, a mixture of CO_2, CO, and CH_4, the formation of pure hydrogen needs additional steps, such as membrane

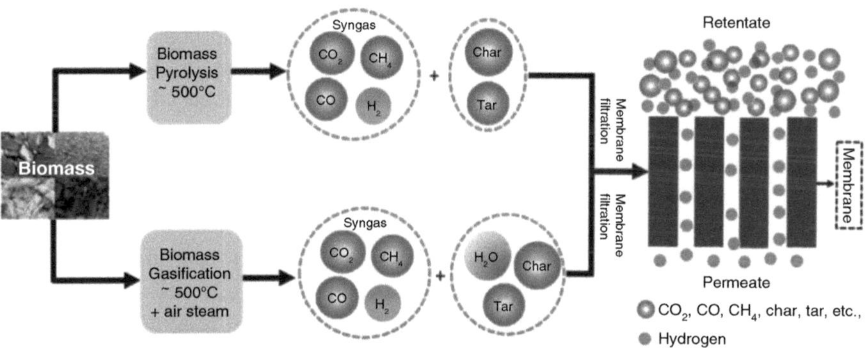

Fig. 5.10 Production of hydrogen from biomass via gasification and subsequent membrane filtration [50]

filtration in certain cases, as depicted in Fig. 5.10 [50]. The term suggests gasification could be a path for hydrogen production and viable in certain cases based on applying product gas for subsequent processes. However, it can also be advised that "gasification a path for direct production of hydrogen" is based on hydrogen production from solid feed in one-step thermal technology, gasification.

Therefore, the scope of hydrogen production from gasification is defined based on applying product gas generated from gasification for cases like integrating gasification systems with solid oxide fuel cell (SOFC) systems, where a certain amount of hydrogen concentration is required for continuously engaging SOFC systems. Hence, our current scientific and industrial community is making many approaches to make gasification technology an emerging platform for hydrogen production in the upcoming sustainable future.

5.3.4 Recent Trends in Demand for Gasification Technology for Industrialization

It is well known that gasification technology has been widely developing worldwide, and the gasification process can handle a broad range of feedstock. However, commercial barriers still exist in implementing the gasification system on a larger scale, and in some cases, limitations also exist in commercial development. In another aspect, gasification technology provides a solution for effectively utilizing renewable sources, i.e., biomass, biosolids, and other sources, generating waste for energy and valuable products in an efficient and socio-economic sustainable manner. Besides those, there are some operational challenges such as reaction temperature, equivalence ratio, and properties of feedstock, which play very critical roles in the process and advancing of gasification technology from lab scale to industrial scale. Additionally, it was also noticed that the government and technology developers play a major role in the commercialization of gasification technology. It can be concluded that some substantial approaches from technology developers can be taken in the direction of minimizing overall costs, optimizing the process, and process automation to commercialize gasification technology. Furthermore, in recent days, techno-economic analysis of gasification technology under the operation of different feedstocks was carried out to identify social and economic feasibility in a large-scale implementation. This analysis for gasification technology depends on my parameters. It includes performance evaluation of a particular designed system for a particular feedstock type, capital cost, operating and maintenance cost, payback period, acceptance of technology, and cost-effectiveness [51].

The continuous rise in energy demand in the current world and the increase in human population, infrastructure, and industries have increased the path for more energy sources. Considering those factors, gasification could substantially facilitate the fulfillment of energy demand. As stated before, gasification could provide an environmental benefit by emitting less polluting gases. On the other hand, some environmental technologies can be developed to capture carbon emissions from low-sulfur transportation fuels. Hence, it can be advised that gasification technology has many undeniable factors/advantages in the aspect of environmental protection with large-scale industrial applications. In addition to this, the continuous advancement of gasification processes and technologies has made it more livable in industries and other energy-driven sectors. Recently, gasification technology has become more attractive due to its flexibility and large-scale implementation, which makes it applicable in different industrial sectors like fertilizer, chemicals, and primary substations of natural gases.

Therefore, in conclusion, the present review of gasification technology indicates that detecting the commercial barrier and making an assessment of the design, development, and evaluation of performance for various types of feedstocks will enable its application and sustainability.

5.4 Fate and Role of Gasification Technology in the Next Generation

The fate and role of gasification technology in emerging technologies and energy demand primarily depend on its usefulness and application in different fields. As schematically shown in Fig. 5.11, the major feedstock is segregated into four types: coal, biomass, petroleum coke, and waste [52]. The waste stream covers many materials, such as biosolids from the water utility, pulp paper waste, agricultural residue, forestry residue, and food-garden organic wastes. In describing the process from feed to used end product in different applications, the gaseous constituent from the gasifier is sent to a gas cleaning system where syngas, a mixture of CO and H_2, and other steam containing pure H_2 [53]. Syngas are used for fuel, the formation of chemicals, transportation fuels, and electric power. The syngas are used in a

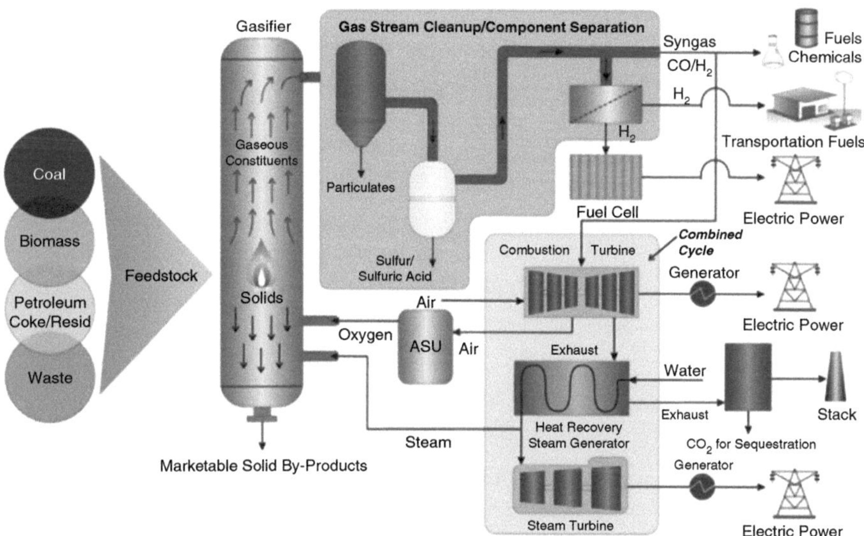

Fig. 5.11 Representation of a gasification process with feedstock flexibility inherent in gasification, as well as products and usefulness of gasification technology [52]

combined cycle mode to generate steam, oxygen, and air through combustion and a turbine to continuously operate the gasification processes. The other aspect is the H_2 generated from separated syngas used for fuel cell operations. Overall, it can be seen that gasification is one of the most cost-effective approaches for generating other valuable products. In contrast to gasification technology, other thermal technologies could provide a more efficient way of energy, but not like gasification, as they have a broad value of integrational approaches in industrial processes.

5.4.1   Sustainable Energy Production from Gasification

The sustainability of the respective technology underlines the sustainable energy production process. In the current world, many renewable sources are available, and energy can be generated using various thermal and non-thermal technologies. The major understanding lies with the liability and visibility of the respective technology [4, 54, 55]. Many questions could be raised before being subjected to any sustainable process.

- Can technology be diversified with other technologies using an integrational approach?
- Can process/technology manage various feedstock with different physical/chemical properties?
- Would technology become techno-economic viable in a larger scale approach?

- Will the technology have a low environmental impact with less emission/release of polluting gases?
- Can we solve some of our ongoing environmental and social problems by approaching this technology?
- Would this technology be able to be part of the circular economy prospect as per government guidelines?

To sum up, few answers to the gasification process question are approachable to understand, whereas some answers would be certain from sustainability from an aspect in an upcoming energy demand in the near future.

5.4.2 Pathway to Obtain Clean and Economical Energy

Pathways to securing clean and economical energy has become an essential requirement for society and the government. Gasification of carbon material is one of the prominent ways of converting the carbonaceous material into high-value syngas cost-effectively and efficiently. Although gasification is not the only way to convert carbonaceous material, some other ways could be effective [55]. However, the flexibility of operating a gasification process, design of the gasifying reactor, optimization of operating parameters, and scaling of the gasification process to an industrial/commercialized scale have made us adopt this technology. In the past few decades, the drive for gasification was primarily focused on converting coal into energy and valuable products [56]. Then biomass received major attention in converting higher value energy and material through the process of gasification as well as in terms of converting the renewable source to energy. Now and in the near future, it has been predicted that a wide range of feedstocks (biosolids, municipal solid waste, animal waste, agricultural waste) will be handled through gasification. This will convert renewable sources and waste into energy (Fig. 5.12).

Consecutive ness in converting different ranges of feed/carbonaceous material for gasification would have solidified the path by stating that the gasification process/technology is a clean route for effectively generating energy and valuable products.

5.4.3 Commercial Opportunity in the Utilization of Waste for Energy Generation

In recent days, the rate of generation of different waste materials has become increasingly saturated because of the continuous rise in population. The waste is generated from various sources, classified into potential solid waste streams (fly ash, mine waste, textile waste, paper & cardboard, plastic waste, waste timber, used solar panels, e-waste, and C&D waste). The other classification is agricultural and

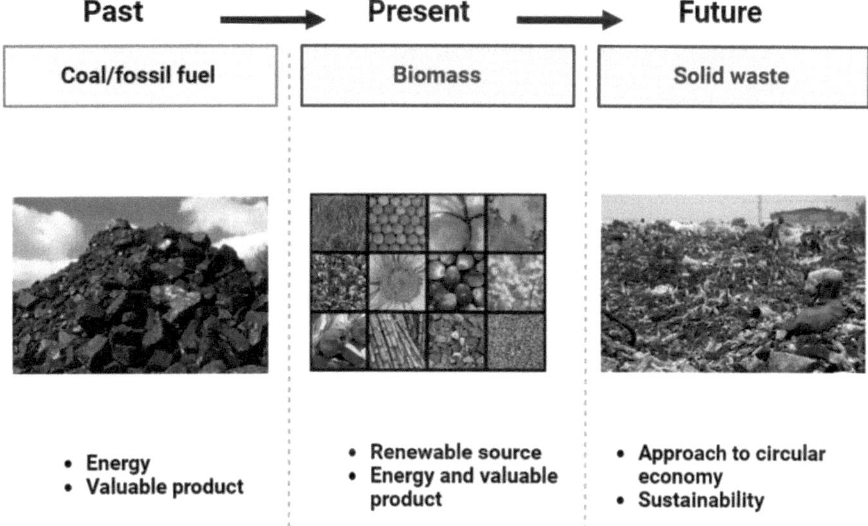

Fig. 5.12 Progressive observations of past and present with a futuristic vision for gasification

domestic waste (garden waste, food waste, municipal solid waste, compost rejects, crop residue, aquatic biomass, cow dung, paper sludge, and chicken manure). It was stated that three wide ranges of feedstocks available, which were considered waste apart from fossil fuel and biomass. In favor of the wide range of feedstocks, a proper design of the gasifying reactor, development of an optimized process, and all requirement considerations of the pre-and post-gasification process would enable the future potential of commercialization of gasification technology in waste management [57, 58]. However, certain constraints and challenges needs to be solved, as they are well defined by government policy and society. Challenges and steps must be considered before performing a commercialization approach, as presented below.

- Selection and consideration of appropriate feedstock based on government policy and environmental regulation.
- Consideration of the level of contamination and whether it would be possible to reduce it to the desired level.
- To continue the gasification process, we could get waste feed at level capacity.
- Will it be possible to perform the gasification process using a wide range of feedstock (waste samples) with varying physical and chemical properties?
- Cost and time for operating a gasification process should be bound to optimized calculation.
- Easy accessibility to the desired waste feedstock at the optimized cost of cleaning and segregation of undesired elements.

5.5   Role and Responsibilities of Scientific Communities for Gasification Technology Protection

Meanwhile, the continuous development of gasification technology and a wide range of feedstock diversification have put the scientific community under responsibilities and duties. In the current global competition for new technologies and processes, the scientific communities need to do additional research on certain key objectives to keep the previously developed in the same race. The three major key objectives that ascertain the role and responsibilities of scientific communities are listed below. A brief discussion on each is presented in the following sections.

- Enhancement of gasification technology and its economic efficiency.
- Broad analysis of possible feedstocks for the sustainability of gasification technology.
- Approach to variability and investigation of gasifying environments for meeting the zero greenhouse gas emissions.

5.5.1   *Enhancement of Gasification Technology and its Economic Efficiency*

In this section, a possible set of solutions is outlined for stepping the gasification technology and its efficiency. The enhancement of gasification technology is a multi-directional approach where the scientific approach can be addressed from different directions in developing building blocks for a sustainable future [59]. Those scientific approaches include understanding the gasification mechanisms by prioritizing the reaction pathways and kinetics in conditions with different feedstocks. Also, attention must be paid to designing suitable reactors, optimizing processes, and other operating conditions. The reactor design should consider destructing the contaminated feedstock such as biosolids and other waste, and consideration must also be made to the low emission of polluting gases into the environment [6]. Finally, an investigation should be made into commercializing gasification technology by following government policy and guidelines, as detailed in a pyramidal structure in Fig. 5.13.

The economic efficiency of gasification technology primarily depends on the cost associated with the technology in scaling up compared to other possible/competing technologies. Also, efficiency can be determined based on how the gasification process is demanded over other approachable technologies [4]. For example, a certain number of these can be approachable in the thermal and non-thermal aspects for converting waste to energy. The key question is whether the gasification process is more economical and efficient among the technologies available. If it is not economically efficient to the expected levels, the scientific and industrial community needs to act together for further consideration and evaluation strategies. Therefore, economic efficiency needs to focus on cost, process, and time efficiency with optimal yield of desired product and energy.

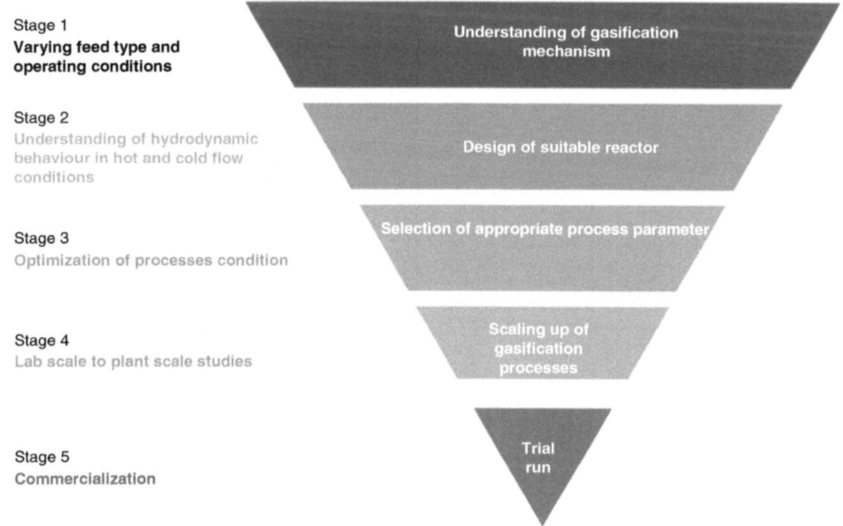

Fig. 5.13 Key processes/pathways for the commercialization of gasification technology via government policies and guidelines

5.5.2 Broad Analysis of Possible Feedstocks for the Sustainability of Gasification Technology

The continuous depletion of solid fossil fuels due to a rise in population with increased energy need empowered the other sources to be past of solid fuel in a gasification process. In other words, the rise in population has caused a series of environmental problems. As a technology provider, the concerns are not only about forming valuable products and generating energy but also about solving environmental problems with technology [6, 60]. In the current scenario of environmental problems, effective solid waste management would be an approach to making the gasification process sustainable. Solid wastes are generated from many different sources. Another key challenge is associated with the effectiveness of the gasification process to manage waste to energy in line with government policies and guidelines. This is because the classification of waste based on different localities varies with the respective government rules. Based on all classifications, all waste collected from different sources is broadly classified into two major categories. One category is potential solid waste streams from different sources, excluding domestic and agricultural waste. The other category is waste generated from domestic and agricultural sources. A pictorial representation of these classified wastes is shown in Fig. 5.14.

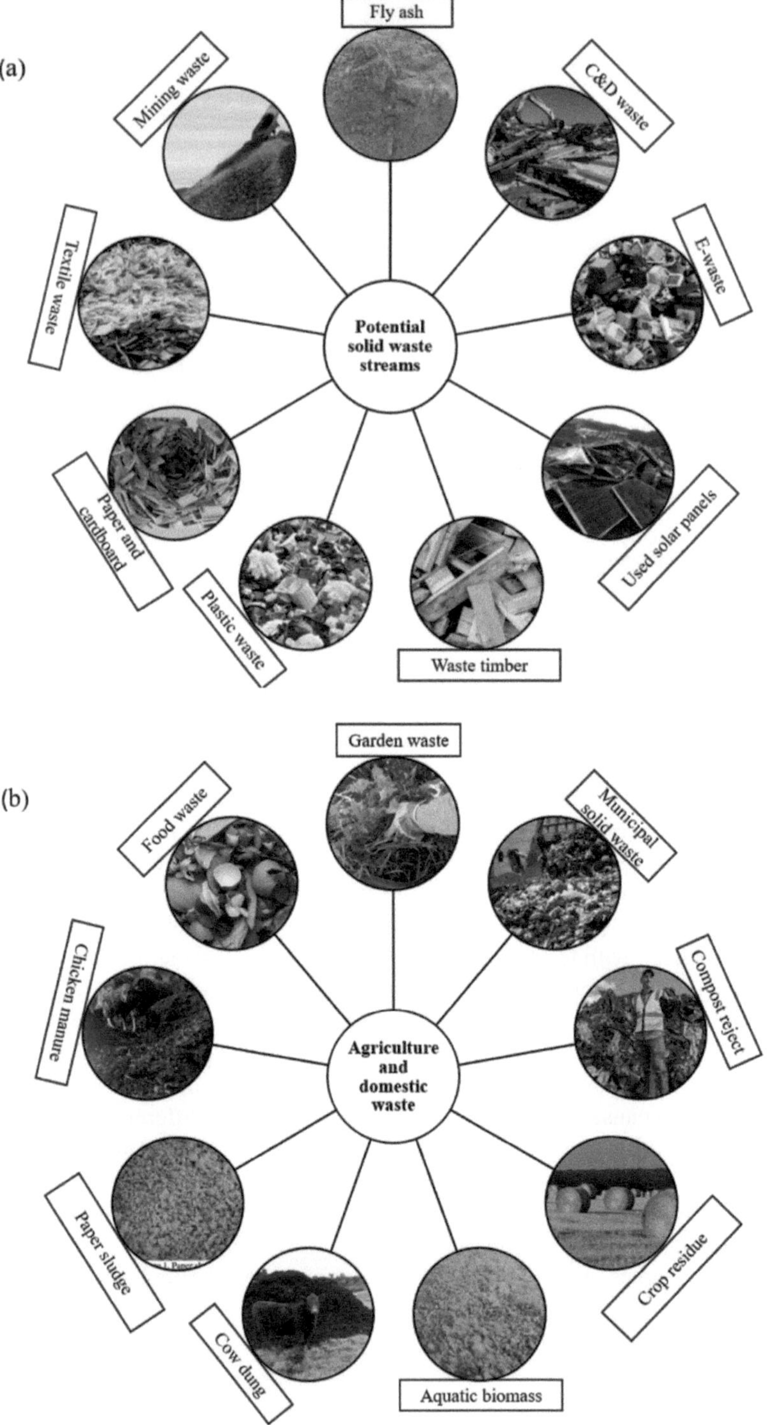

Fig. 5.14 Potential solid waste sources (**a**) and agricultural & domestic waste (**b**) for gasification [61]

5.5.3   Approach to Variability and Investigation of Gasifying Environments for Zero Greenhouse Gas Emissions

The greenhouse gas mainly consists of carbon dioxide, methane, ozone, nitrous oxide, chloroform carbon, and water vapor. The gasification process usually does not participate in the release of all gases. However, in some situations/particular cases, green gases like CO_2 are released into the atmosphere. In comparison to combustion and other thermal oxidation processes, the gasification of solid fuel underlying the partial oxidation mechanism is a possibility that is less likely to form and release CO_2 into the atmosphere [62, 63]. In arguing the definition of the gasification process with limited formation CO_2, the real lab and industrial gasification operations are not nearly the same. Many internal and external factors could potentially affect the gasification process, resulting in a deviation from the true gasification definition mechanisms. Those include feed size, gasifying agent, operating temperature, type reactor, and source of heat supply. Based on the required composition of the product and other efficiency factors, the investigation should be made very less in the level of zero emission of CO_2 to the environment. Based on the understating of the gasification process, here are some of the steps that assist in reducing CO_2 formation.

- It may benefit a mixed gasifying environment instead of a single gasifying one.
- Operation at high temperatures is beneficial for a higher rate of gasification, but it may adversely affect the formation of more CO_2.
- Higher catalytic effect of inherent alkali and alkaline carth metal (AAEM) species could benefit catalyze the process, but in certain cases, it may affect product gas formation with more formation of CO_2.
- The particle size needs to be optimized so the intraparticle diffusion limitation can be reduced, avoiding the formation of CO_2.
- The design of the gasification reactor needs to be optimized to avoid the residence time and contact of product gases in the formation of CO_2.

Overall, the future outlook for gasification technology is diversified. Knowing the current growth in science and technology in the aspect of the gasification process would open a big room for further application of the gasification process (Fig. 5.15). These futuristic approaches and applications include the application of solid waste as potential renewable for gasification, integration of gasification and solid oxide fuel cell technology, and production and use of co-gasified biochar energy storage, catalyst, and adsorbent. Also, there are some possible applications as the supply of H_2 in fuel stations for vehicles and the scaling up and commercialization of plasma and supercritical water gasification processes.

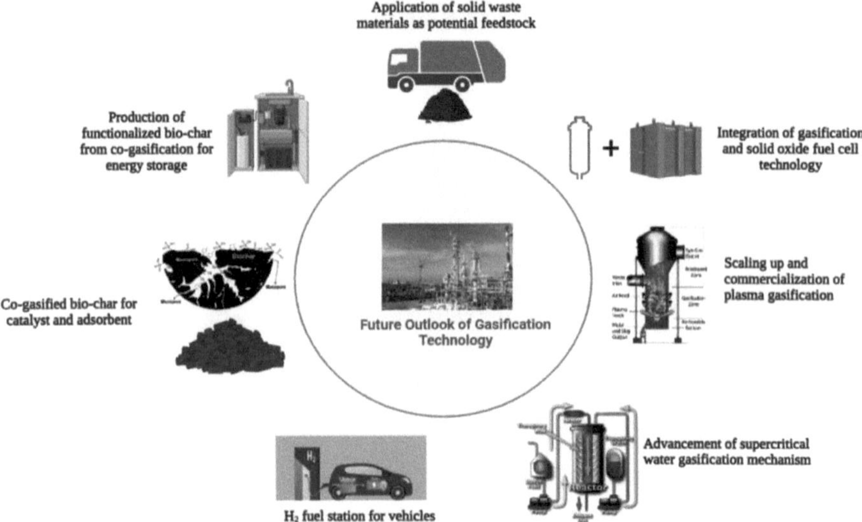

Fig. 5.15 The possible future outlook for gasification technology

References

1. Sajid, M., et al. (2022). Gasification of municipal solid waste: Progress, challenges, and prospects. *Renewable and Sustainable Energy Reviews, 168*, 112815.
2. Janajreh, I., et al. (2021). A review of recent developments and future prospects in gasification systems and their modeling. *Renewable and Sustainable Energy Reviews, 138*, 110505.
3. Erdiwansyah, E., et al. (2023). Analysis of technological developments and potential of biomass gasification as a viable industrial process: A review. *Case Studies in Chemical and Environmental Engineering, 8*, 100439.
4. Pereira, E. G., et al. (2012). Sustainable energy: A review of gasification technologies. *Renewable and Sustainable Energy Reviews, 16*(7), 4753–4762.
5. Ryzhkov, A. F., et al. (2016). Development of entrained-flow gasification technologies in the Asia-Pacific region (review). *Thermal Engineering, 63*(11), 791–801.
6. Hameed, Z., et al. (2021). Gasification of municipal solid waste blends with biomass for energy production and resources recovery: Current status, hybrid technologies and innovative prospects. *Renewable and Sustainable Energy Reviews, 136*, 110375.
7. Santos, S. M., et al. (2023). Waste gasification technologies: A brief overview. *Waste, 1*, 140–165. https://doi.org/10.3390/waste1010011
8. Lopez, G., et al. (2018). Recent advances in the gasification of waste plastics. A critical overview. *Renewable and Sustainable Energy Reviews, 82*, 576–596.
9. Pinna-Hernández, M. G., et al. (2019). Selection of biomass supply for a gasification process in a solar thermal hybrid plant for the production of electricity. *Industrial Crops and Products, 137*, 339–346.
10. Nam, H., & Konishi, S. (2019). Potentiality of biomass-nuclear hybrid system deployment scenario: Techno-economic feasibility perspective in South Korea. *Energy, 175*, 1038–1054.
11. Kan, X., et al. (2017). Energy performance of an integrated bio-and-thermal hybrid system for lignocellulosic biomass waste treatment. *Bioresource Technology, 228*, 77–88.
12. Perna, A., et al. (2018). Performance assessment of a hybrid SOFC/MGT cogeneration power plant fed by syngas from a biomass down-draft gasifier. *Applied Energy, 227*, 80–91.

13. Gómez-Barea, A., Ollero, P., & Leckner, B. (2013). Optimization of char and tar conversion in fluidized bed biomass gasifiers. *Fuel, 103*, 42–52.

14. Iovane, P., Donatelli, A., & Molino, A. (2013). Influence of feeding ratio on steam gasification of palm shells in a rotary kiln pilot plant. Experimental and numerical investigations. *Biomass and Bioenergy, 56*, 423–431.

15. Pandiyan, P., et al. (2022). A comprehensive review of the prospects for rural electrification using stand-alone and hybrid energy technologies. *Sustainable Energy Technologies and Assessments, 52*, 102155.

16. Yao, Z., et al. (2017). Anaerobic digestion and gasification hybrid system for potential energy recovery from yard waste and woody biomass. *Energy, 124*, 133–145.

17. Colpan, C. O., et al. (2010). Effect of gasification agent on the performance of solid oxide fuel cell and biomass gasification systems. *International Journal of Hydrogen Energy, 35*(10), 5001–5009.

18. Digman, B., Joo, H. S., & Kim, D.-S. (2009). Recent progress in gasification/pyrolysis technologies for biomass conversion to energy. *Environmental Progress & Sustainable Energy, 28*(1), 47–51.

19. Pang, S. (2019). Advances in thermochemical conversion of woody biomass to energy, fuels and chemicals. *Biotechnology Advances, 37*(4), 589–597.

20. He, M., et al. (2009). Hydrogen-rich gas from catalytic steam gasification of municipal solid waste (MSW): Influence of catalyst and temperature on yield and product composition. *International Journal of Hydrogen Energy, 34*(1), 195–203.

21. Hu, M., et al. (2015). Hydrogen-rich gas production by the gasification of wet MSW (municipal solid waste) coupled with carbon dioxide capture. *Energy, 90*, 857–863.

22. Tijmensen, M. J. A., et al. (2002). Exploration of the possibilities for production of Fischer Tropsch liquids and power via biomass gasification. *Biomass and Bioenergy, 23*(2), 129–152.

23. Barahmand, Z., & Eikeland, M. S. (2022). A scoping review on environmental, economic, and social impacts of the gasification processes. *Environments, 9*. https://doi.org/10.3390/environments9070092

24. Vaish, B., et al. (2019). Energy recovery potential and environmental impact of gasification for municipal solid waste. *Biofuels, 10*(1), 87–100.

25. Ramos, A., & Rouboa, A. (2022). Life cycle thinking of plasma gasification as a waste-to-energy tool: Review on environmental, economic and social aspects. *Renewable and Sustainable Energy Reviews, 153*, 111762.

26. Chanthakett, A., et al. (2021). Performance assessment of gasification reactors for sustainable management of municipal solid waste. *Journal of Environmental Management, 291*, 112661.

27. Ahrenfeldt, J., et al. (2013). Biomass gasification cogeneration—A review of state of the art technology and near future perspectives. *Applied Thermal Engineering, 50*(2), 1407–1417.

28. Molino, A., Chianese, S., & Musmarra, D. (2016). Biomass gasification technology: The state of the art overview. *Journal of Energy Chemistry, 25*(1), 10–25.

29. Khandelwal, K., et al. (2024). Hydrogen production from supercritical water gasification of canola residues. *International Journal of Hydrogen Energy, 49*, 1518–1527.

30. Nanda, S., et al. (2018). Catalytic gasification of wheat straw in hot compressed (subcritical and supercritical) water for hydrogen production. *Energy Science & Engineering, 6*(5), 448–459.

31. Nanda, S., et al. (2017). An assessment of pinecone gasification in subcritical, near-critical and supercritical water. *Fuel Processing Technology, 168*, 84–96.

32. Leong, Y. K., et al. (2021). Supercritical water gasification (SCWG) as a potential tool for the valorization of phycoremediation-derived waste algal biomass for biofuel generation. *Journal of Hazardous Materials, 418*, 126278.

33. Demirel, E., Erkey, C., & Ayas, N. (2021). Supercritical water gasification of fruit pulp for hydrogen production: Effect of reaction parameters. *The Journal of Supercritical Fluids, 177*, 105329.

34. Adar, E., Ince, M., & Bilgili, M. S. (2020). Supercritical water gasification of sewage sludge by continuous flow tubular reactor: A pilot scale study. *Chemical Engineering Journal, 391*, 123499.
35. Brian, P., et al. (2019). Gasification kinetics in continuous supercritical water reactors. In P. Igor (Ed.), *Advanced supercritical fluids technologies*. IntechOpen. p. Ch. 5.
36. Guo, L., & Jin, H. (2013). Boiling coal in water: Hydrogen production and power generation system with zero net CO2 emission based on coal and supercritical water gasification. *International Journal of Hydrogen Energy, 38*(29), 12953–12967.
37. Chen, J., et al. (2021). Process in supercritical water gasification of coal: A review of fundamentals, mechanisms, catalysts and element transformation. *Energy Conversion and Management, 237*, 114122.
38. Salimi, M., et al. (2016). Hydrothermal gasification of different agricultural wastes in supercritical water media for hydrogen production: A comparative study. *International Journal of Industrial Chemistry, 7*(3), 277–285.
39. Zhang, Y., et al. (2020). Simulation of microwave-assisted gasification of biomass: A review. *Renewable Energy, 154*, 488–496.
40. Li, J., et al. (2021). Review of microwave-based treatments of biomass gasification tar. *Renewable and Sustainable Energy Reviews, 150*, 111510.
41. Arpia, A. A., et al. (2022). Microwave-assisted gasification of biomass for sustainable and energy-efficient biohydrogen and biosyngas production: A state-of-the-art review. *Chemosphere, 287*, 132014.
42. Achinas, S. (2020). An overview of the technological applicability of plasma gasification process. In P. Singh, R. P. Singh, & V. Srivastava (Eds.), *Contemporary environmental issues and challenges in era of climate change* (pp. 261–275). Springer.
43. Oliveira, M., et al. (2022). A review on plasma gasification of solid residues: Recent advances and developments. *Energies, 15*(4), 1475.
44. Chu, C., et al. (2024). Mechanistic exploration of polytetrafluoroethylene thermal plasma gasification through multiscale simulation coupled with experimental validation. *Nature Communications, 15*(1), 1654.
45. Mankasem, J., Prasertcharoensuk, P., & Phan, A. N. (2024). Intensification of two-stage biomass gasification for hydrogen production. *International Journal of Hydrogen Energy, 49*, 189–202.
46. Chanthakett, A., et al. (2024). Chapter 4 – Hydrogen production from municipal solid waste using gasification method. In M. M. K. Khan, A. K. Azad, & A. M. T. Oo (Eds.), *Hydrogen energy conversion and management* (pp. 103–131). Elsevier.
47. Sheth, P. N., & Babu, B. V. (2010). Production of hydrogen energy through biomass (waste wood) gasification. *International Journal of Hydrogen Energy, 35*(19), 10803–10810.
48. Zhang, Y., et al. (2019). Hydrogen production through biomass gasification in supercritical water: A review from exergy aspect. *International Journal of Hydrogen Energy, 44*(30), 15727–15736.
49. Singh, A., Shivapuji, A. M., & Dasappa, S. (2023). Hydrogen production through agro-residue gasification and adsorptive separation. *Applied Thermal Engineering, 234*, 121247.
50. Besha, A. T., et al. (2020). Deployable membrane-based energy technologies: The Ethiopian prospect. *Sustainability, 12*(21), 8792.
51. Sansaniwal, S. K., et al. (2017). Recent advances in the development of biomass gasification technology: A comprehensive review. *Renewable and Sustainable Energy Reviews, 72*, 363–384.
52. *Gasification introduction*. National Energy Technology Laboratory.
53. Mantulet, G., Bidaud, A., & Mima, S. (2020). The role of biomass gasification and methanisation in the decarbonisation strategies. *Energy, 193*, 116737.
54. Sobamowo, G. M., & Ojolo, S. J. (2018). Techno-economic analysis of biomass energy utilization through gasification technology for sustainable energy production and economic development in Nigeria. *Journal of Energy, 2018*, 4860252.

55. Alptekin, F. M., & Celiktas, M. S. (2022). Review on catalytic biomass gasification for hydrogen production as a sustainable energy form and social, technological, economic, environmental, and political analysis of catalysts. *ACS Omega, 7*(29), 24918–24941.
56. Hosseini, S. E., & Wahid, M. A. (2016). Hydrogen production from renewable and sustainable energy resources: Promising green energy carrier for clean development. *Renewable and Sustainable Energy Reviews, 57*, 850–866.
57. Moya, D., et al. (2017). Municipal solid waste as a valuable renewable energy resource: A worldwide opportunity of energy recovery by using waste-to-energy technologies. *Energy Procedia, 134*, 286–295.
58. Guran, S. (2018). Chapter 8 – Sustainable waste-to-energy technologies: Gasification and pyrolysis. In T. A. Trabold & C. W. Babbitt (Eds.), *Sustainable food waste-to-energy systems* (pp. 141–158). Academic Press.
59. Santos, M. P. S., & Hanak, D. P. (2022). Techno-economic feasibility assessment of sorption enhanced gasification of municipal solid waste for hydrogen production. *International Journal of Hydrogen Energy, 47*(10), 6586–6604.
60. Wu, C. Z., et al. (2002). An economic analysis of biomass gasification and power generation in China. *Bioresource Technology, 83*(1), 65–70.
61. News, D. R., *We are so bad at composting, whole truckloads are rejected and sent to the dump.*
62. Yang, Q., et al. (2018). Hybrid life-cycle assessment for energy consumption and greenhouse gas emissions of a typical biomass gasification power plant in China. *Journal of Cleaner Production, 205*, 661–671.
63. Man, Y., et al. (2014). Environmental impact and techno-economic analysis of the coal gasification process with/without CO2 capture. *Journal of Cleaner Production, 71*, 59–66.

Chapter 6
Conclusions and Recommendations

Embrace Innovation, Prioritize Sustainability, and Seek Improvement for Brighter Future

6.1 Introduction

The chapters of this book are generally intended to cover all aspects of gasification technology. This includes the significance and importance of every type of gasification technology and the respective advantages over one another by discussing their applications. In addition, the fundamental aspect of every type of gasification technology and its mechanisms are explained. These insights into gasification technology and its wide implications for the variety of feedstocks are compared with one another. Additionally, the challenges and benefits associated with each are presented by reviewing the research and industrial engagement in the past and present. Overall, future expectations and trends that need to be done are also elaborated. The above-mentioned steps and paths are subject to each individual chapter and are specific to those particular intended contents. In alignment with those paths and subject to the content of each individual chapter, overall conclusions and recommendations will be presented by surveying the book. The points stated below are some conclusions that have been taken based on the current situation/place of gasification technology in an emerging energy-demanding and fast-growing world. The conclusions presented here are based on covering the fundamental and engineering applications of technology with socio-economic implications.

6.2 Conclusions

Regardless of any type of gasification reactor/process, understanding the mechanism of char/bio-char gasification is very important as it is considered as the rate-limiting/controlling step for the overall gasification process. To understand the mechanism of the rate-limiting step, a thorough knowledge of reaction kinetics and

© The Author(s), under exclusive license to Springer Nature Switzerland AG 2024
M. K. Jena, H. B. Vuthaluru, *Gasification Technology*,
https://doi.org/10.1007/978-3-031-71044-5_6

proper kinetic model is required. Based on the current knowledge, no proper kinetics models have been developed so far, as kinetic parameters keep changing during gasification. This is because some of the kinetic models are developed based on initial changes in char/bio-char properties, and it was noted that the change in char properties during the gasification processes have a significant impact on the respective gasification rates. This indicates that a proper understanding of gasification is largely dependent on the reaction, and this leads to the adoption of the kinetic compensation effects (KCE) mechanism because the gasification reactions is a solid-gas heterogeneous reaction. The KCE studies would provide a platform for interpreting the variation of kinetic parameters such as apparent activation energy (E_{app}) and apparent frequency factor (lnA_{app}) during the progress of the gasification process.

In optimizing the product yield, understanding the path of product gas formation in the process of gasification is very important. It has also been concluded from past studies that the varying particle size leads to some possibilities for changing the path of product gas formation, which subsequently affects the desired product gas yield in product gas concentration. In addition to optimizing the gasifying agent, optimizing solid fuel particle size for gasification is very important. Future work at lab- and plant-scales needs to focus on particle size with respect to differences in solid fuel composition and properties. Further analysis also indicates that the alkali and alkaline earth metal (AAEM) species significantly catalyze the gasification process. However, the catalytic role of those AAEM species depends on the chemical form of inherent AAEM species gained during gasification.

The design of a gasification reactor from the lab scale studies is not a single-step approach. The pathway includes generating data from a lab-scale reactor, synthesizing rate law from the data, and developing a reaction mechanism with rate-limiting steps. Once those have been derived, the next approach will be given to estimating rate law parameters and the design of the gasification reactor. Fundamentally, those steps are developed from a basic understanding of reaction engineering, fluid mechanics, and thermodynamics. However, all this knowledge will combine to develop a gasification technology on a lab scale and scale it up from bench to plant scale. The fate and role of the gasification environment are primarily defined by its impact on industrial applications. This book has explained each class of gasification technology in a very detailed manner. However, certain challenges are associated with every kind of gasification technology; more attention must be given in eradicating that barrier and making gasification technology promising and sustainable.

In some cases, gasification of char/biochar in a mixed gasifying environment provides additional benefits to increase the product yield. To design a mixed gasifying environment, some considerations need to be taken for the controlled regime, as a variation in the kinetics-controlled regime to the diffusion-controlled regime occurs with the change in the gasifying environment. Some scenarios would be preferred in overlapping one regime specific to a particular gasifying environment with another. This leads to the overlapping of reaction pathways in relation to one another and creates a possibility of a higher yield of the desired product in the gas composition.

In the current scenario, a wide range of feedstocks can be used as a solid fuel for gasification. Some other sources, including fossil fuel, are related to biomass, biosolids, municipal solid waste, food waste, and other renewable waste sources. However, adopting any kind of solid feed to commercialize plant scale from lab scale needs to follow some sort of guidelines and policies as defined by the respective regional government and other environmental controlling authorities. Therefore, identifying the potential feedstocks for a commercialized gasification plant is very necessary.

In the current development of the new approach to the gasification process, viable techniques such as supercritical water gasification, plasma gasification, and others are well enough to solve problems associated with previous gasifiers. However, approaches should include hydrodynamic modeling and process modeling so that these can be scaled up to industrial-scale gasifiers and easily installed or added to integrated process systems.

In the context of developing a process dynamic model of any kind of gasifier in hot conditions, including appropriate kinetic parameters/kinetic model in the hydrodynamic model is necessary. These sets of kinetic parameters and/or kinetic models can be obtained by thoroughly understanding the mechanisms of solid-gas heterogenous gasification reactions with the insight of kinetic compensation effects. Similarly, for the development of a process model with the inclusion of the gasification process in an integrated system, efforts must be made to select the right model for the gasification reactor.

Overall, comparing past and present efforts on gasification technology, an ample amount of research has been given to the development of the gasification process with technique; similarly also, equal attention was given to applications of other renewable sources as feedstock and/or co-feedstock for gasification. It can be stated that the gasification process is one of the promising technologies for the thermochemical conversion of solid fuel that led to making the gasification process more efficient and sustainable in the current energy-demanding world. Also, equal importance was given to making the gasification process a part of an integrated process system. This approach has created a platform for the gasification process to be a more reliable technology/process in many industrial applications, including the existing benefits.

6.3   Recommendations

The development of technology and advanced insights into science is a continuous process. Gasification technology can be considered a promising technology for thermochemical conversion of solid fuel in a wide range of solid fuel properties. At the same time, some consideration still needs to be given to some aspects of gasification technology to make it more competent among other thermal technologies in the future. Here are some major highlights that can be implemented in future work to make the gasification process more sustainable.

- An approach should be given to recover more heat from product gas in an integrated process system.
- The gasifying agent should be adequately optimized so that it is easily accessible and cost-effective and should be approachable to generate the desired product gas yield.
- In developing the hydrodynamic models, the impact of particle size distribution and hydrodynamic conditions relating to intraparticle and interparticle diffusion limitations affecting the path of product gas formation require further attention.
- Understanding reaction kinetics should be done to gain better insights into the gasification mechanisms by interpreting the solid-gas heterogeneous mechanism through kinetic compensation effect studies.
- In the development of gasification technology, particularly for certain kinds of feedstock (municipal solid waste, biosolids, food waste, etc.), the inherent nature and physiochemical properties of solid fuel need to be considered properly. This would enable the design and operation of appropriate frontend and backend operations for the safer operation of the gasification process in an interpreted process plant.
- In some industrial scenarios, gasification at higher temperatures was preferred due to the high rate of gasification reaction. However, it will be beneficial to understand the pathways for the formation key product gaseous species This can be achieved by properly specifying the controlled regime and the respective intrinsic reaction kinetics.
- The commercialization of any form of gasification technology should start at a lab scale with a feed handling capacity of a milligram to understand the properties and kinetics and then scale up to the gram in a lab-scale reactor to visualize the hydrodynamic behaviors and product yields. Finally, the process/technology should be scaled up to a plant/semi-pilot scale from a lab-scale for industrial application.
- In developing gasification technology, attention must be given to the required approach to observe the subsequent effect of developed gasification technology, which was schematically shown in Fig. 6.1.

Overall, all the suggested recommendations are approachable to limited conditions based on the type of solid fuel and gasification process. Based on the advancement of technology, it could vary with circumstances.

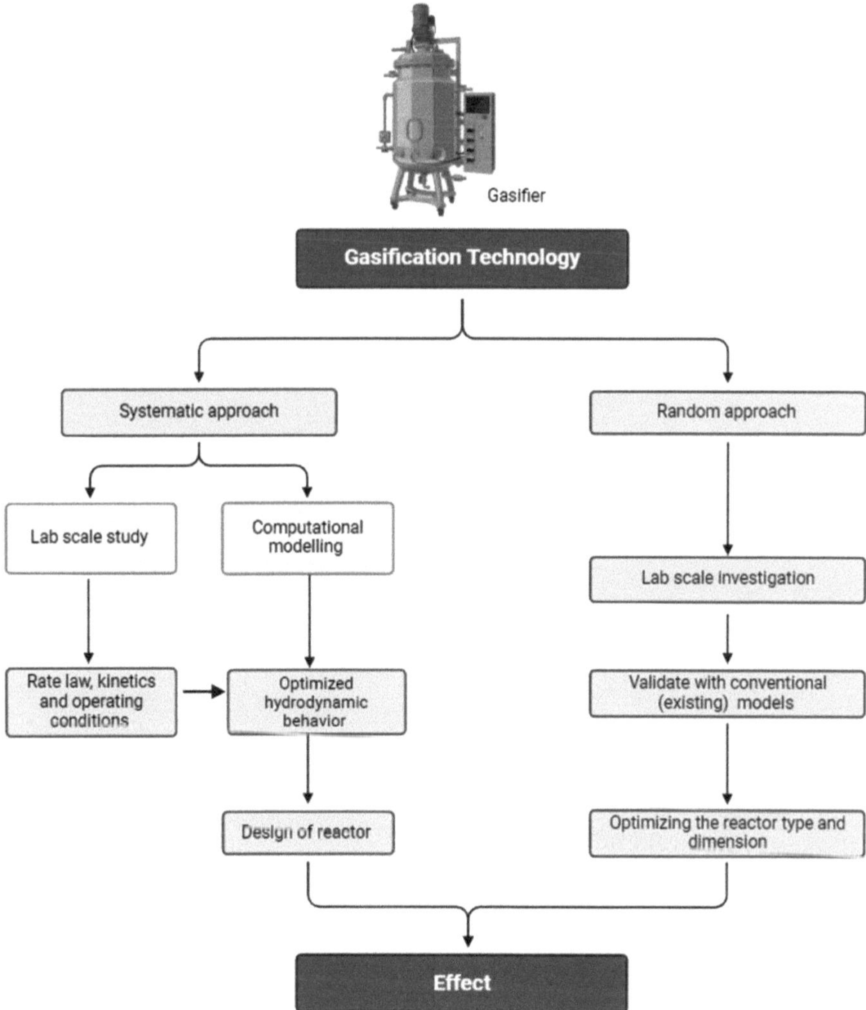

Fig. 6.1 Schematic pathway for the development of gasification technology through systematic and random approach

Index